Ethics & Ecosystems

CIS Policy Monographs 29

Ethics & Ecosystems

PROTECTING HUMAN INTERESTS AND ENVIRONMENTAL VALUES

Barry Maley

THE CENTRE FOR INDEPENDENT STUDIES

1994

Published December 1994 by

The Centre for Independent Studies Limited

National Library of Australia

Cataloguing-in-Publication Data:

Maley, Barry, 1925- .
Ethics and ecosystems.

Bibliography.
Includes index.
ISBN 1 86432 002 8.

1. Environmental ethics. 2. Environmental policy – Moral and ethical aspects. I. Centre for Independent Studies (Australia). II. Title. (Series: CIS policy monographs; 29).

179.1

Book Design by Heather M. Clements.
Printed by CPN Publications, Fyshwick, ACT.
Typeset in Garamond 10pt and Garamond 3.

Contents

PART TWO
The Preservationist Ethic & Western Civilisation

PART THREE
Implementing the 'Mainstream' Ethic

Foreword

In the rush of day-to-day politics there is little time to consider underlying assumptions or to carefully check facts. The long-term consequences of not stepping back to consider the broader picture can be very serious. This is the case with the environmental debate today. For example, Australia has committed itself to reducing greenhouse gas emissions; a project that will impose considerable costs on consumers, companies and the community. Environmental groups would like the government to go even further by imposing a carbon tax on greenhouse gas producers.

As Barry Maley shows in *Ethics and Ecosystems* many of the claims about greenhouse gas and other environmental problems are unproven and unlikely to be true. Perceptions of an impending environmental calamity are often the product of wild exaggeration and distortion by some of the less scrupulous sections of the environmental movement. The issues raised by this go beyond considerations of the ethics of political activism. It is necessary to examine why environmentalists hold the views they do; why they feel that their cause is so important that misleading and deceiving the public on environmental matters is a justifiable tactic.

Barry Maley traces the influence on environmental thought of the idea that nature possesses intrinsic value, independent of human values or interests. When the environment is seen as having intrinsic value, stopping its use by humans – even at great cost to them – is taken to be justified as a policy goal. Important human interests and truth end up being acceptable casualties in the battle to protect the environment's intrinsic value. In order to get to the base of environmentalism as a political problem, the fundamental assumption that nature has intrinsic value needs to be challenged. *Ethics and Ecosystems* shows the implausibility of the view that nature has intrinsic value, and proposes instead that the only measure of the value of nature and the environment are the consequences for human spiritual and practical interests of using, or preserving, natural things. Such a morality does not mean that environmental issues should be neglected, or that there are not important moral issues relevant to our treatment of other species. It does mean that these issues are put in the context of human values and interests.

The task of environmental policy is to develop procedures capable of taking account of the many interests and values that are at stake. *Ethics and Ecosystems* is rightly critical of the 'command-and-control' approach favoured by governments and environmentalists. Instead, it advocates a 'self-organising' environmental policy; one that creates appropriate incentives for environmental preservation. Property rights and a market economy, within a generalised framework of 'environmental common law', can create incentives to care for the environment and to seek substitutes for scarce resources. Ironically, the ideological hostility of important sections of the environmental movement to these 'capitalist' institutions is a major obstacle to on-going effective protection of the environment.

Greg Lindsay
Executive Director

About the Author

Barry Maley has been Senior Fellow with the Centre for Independent Studies since 1989. Prior to taking up this appointment he was Senior Lecturer in Behavioural Science at the University of New South Wales.

He was educated at the University of Sydney and the Australian National University. He has held appointments as Visiting Scholar at Oxford and Cambridge Universities, the University of California, Bath University, and Research Fellow at the Institute of Commonwealth Studies, London.

Since joining the Centre for Independent Studies, he has been working mainly on welfare and family policy, environmental policy and aspects of industrial training policy. Barry has published a number of articles on such matters in the Centre's journal *Policy*, together with two books and a number of articles in newspapers and academic journals. His most recent publication for the Centre for Independent Studies was *Marriage, Divorce and Family Justice* (1993).

Author's Preface

This study is concerned with moral and ethical issues arising from the environmental debate, and with developing an approach to environmental policy that takes these ethical concerns seriously within an economically and socially responsible framework.

The first part of the study deals with general questions of environmental values, animal rights, and the responsibilities of mankind to other species. The second part locates ethical questions about environmental matters within the secular and religious traditions of Western civilisation. Then, in tracing recent developments, I examine the melding of an anti-Western ideology with techniques of public deception in relation to some key issues of the 'environmental crisis'. The third part is ethical in the broader sense in examining the 'ways of life', within the mainstream Western traditions, that might best reconcile environmental protection and species preservation with liberty and abundance.

There is little doubt that most Australians today are deeply concerned about protecting the natural environment and native species. That concern has brought sharply to attention a whole range of problems which, say, thirty years ago, interested only a small vanguard.

For example, the preservation of 'biodiversity', which includes all species from the largest animals and plants down to insects, fungi and microbes, has assumed growing importance as an environmental issue. Indeed, it has become the cutting edge of the environmental movement. Not only does it subsume most of the issues at stake in protecting the natural environment, it introduces a new range of practical and moral problems peculiar to itself.

Mined land, eroded landscapes, degraded forests, polluted air, land, sea and rivers, can usually be rehabilitated to a greater or lesser degree or protected from further decay. But a species once lost is lost for ever, although the science of preserving genetic material and perhaps re-animating 'extinct' species from such material is advancing rapidly. At present, however, there is a finality and irretrievability about the disappearance of a species which elicits powerful human emotions; and, similarly, if a species is seen to be in danger of extinction. This is especially true of the more appealing, or 'charismatic' species, such as whales and koalas. The call to preserve species therefore evokes a special kind of response that greatly empowers preservation movements and legitimises government action aimed at preservation. If anything, Australians take a special pride in the uniqueness of their native species, so their responses to threatened extinction may well be stronger on that account.

Moreover, since it is possible that a great many habitats may be found to contain one or more species unique to it, the scope for preserving such environments, with consequences for their economic use and availability, is extended.

Also important in debates about the preservation of biodiversity is the contention that species are **valuable in themselves** (quite apart from any human needs, attitudes or desires). A philosophically-related argument, but one not necessarily employed by those who wish to preserve biodiversity, is that at least some species such as the higher mammals – have **rights** of their own which are comparable to human rights and which deserve respect and acknowledgment on their own account. Both of these positions are aspects of a 'biocentric' (or 'non-anthropocentric') ethics and are usually opposed to what has been called 'anthropocentric' ethics.

'Anthropocentric' ethics comprises the mankind-centred arguments that defend preservation of species on the grounds that to do so serves a variety of human values and interests, both material and spiritual. This ethic may be expressed as a practical or utilitarian interest in preserving species that might be useful to man in various ways. Domesticated animals and plants are an obvious example. But this may be extended to include various other plants and organisms that are essential in agriculture, pharmaceuticals, animal husbandry, and so on. Since the future use value of both known and unknown species is indeterminate, this provides a practical argument for their indefinite preservation, quite apart from other considerations. Accordingly, on this argument, there may be millions of insects, plants and micro organisms, as yet unidentified and unclassified, that could be useful in these ways, so it would be wise to preserve the habitats in which they now exist, such as Brazilian or North Queensland rain forests.

But the anthropocentric ethic may also be expressed in a human interest in preservation for less 'practical' reasons. There is special interest, among sizeable sections of the population, in preserving species for purely spiritual or aesthetic reasons because their mere existence, the mere contemplation of the richness of the world and the presence of species in it, is deeply satisfying. In more extreme forms, this may take an openly religious form as reverence for the natural world and its creatures as 'God's Creation' and as sacred on that account.

In sum, there are various arguments and strong emotions behind the demand for preservation of biodiversity and the protection of environmental amenity in general. Together, they raise moral and practical issues of great importance; issues that acutely confront existing attitudes and ongoing industrial and agricultural activities that depend upon use of the land and its resources in ways that may threaten the existence of species. In its simplest forms, the ethic of preservation values stasis and absence of change, while the ethic of development values dynamism and transformation, and, in some debates, extreme versions of these competing ethics have become irreconcilable. But for non-extremists on either side, there is hope that the objectives of both may be substantially achieved by means already at hand, even though there may be honest disagreements about the emphasis on one means or another.

The reconciliation of species preservation and environmental protection with the demands of a civilisation dedicated to dynamic transformations and development in order to maintain its institutions and ways of life therefore requires of those same institutions a corresponding ingenuity and responsiveness in order to resolve the dilemmas which are being raised.

The dilemma is aptly summed up, if not solved, by the formula which has gained currency in the debate which has been engaged. This is the notion of 'ecologically sustainable development'. The phrase suggests that the natural environment and its resources must be used and managed in such a way that they will not be irreversibly depleted or degraded, or habitats threatened, while providing a constant, or increasing, supply of materials for the indefinite future, so that future generations will not suffer because of the selfish depredations of the present generation. In the pages that follow, I discuss what kinds of things constitute 'resources' and suggest that the answer is by no means a straightforward one.

PART ONE

Clearing the Ground – Some Issues in Ethics & Moral Philosophy

Introduction to Part One

In the following five chapters which comprise Part 1, I discuss some of the fundamental ethical-philosophical questions raised by the opposition between 'anthropocentric' ethics and 'non-anthropocentric' or 'biocentric' or 'ecocentric' ethics.

I believe one of the central questions to be answered is whether there can ever be a system of ethics that justifies a special and unique place, and special consideration for natural objects and natural processes, including species other than man. This is a burning issue for many who specialise in pondering how mankind should conduct itself in relation to the natural world and other species .

One view would argue that ethics only has one valid subject matter; and that is the moral relationships among men. Questions of how we treat the environment and other creatures are relevant only insofar as that treatment has consequences for *human values and human interests*. An alternative view argues that natural things and other creatures demand ethical consideration *in their own right,* quite independently of any consideration of human values and human interests.

The conclusions emerging from an examination of such matters bear very directly upon questions of environmental policy and the ethical stances that justify such policies. It helps us, for example, to deal with questions like the following:

(i) Do animals have rights?

(ii) What sorts of moral duties do we have to other species?

(iii) Do trees have moral standing?

(iv) Are the higher mammals our moral equals?

(v) Are we *obliged* to deal with the natural world in certain ways?

(vi) When, if ever, may we kill or destroy other creatures with a clear conscience?

It is clear that a well-considered set of policies cannot ignore the answers to such questions and yet be held to be reasonably comprehensive in dealing with the ethical issues that our relations with the environment and other species raise. They are not easy questions to answer. In one form or another they have occupied mankind for thousands of years; but they cannot be avoided.

CHAPTER 1

Ethics & Environmental Issues

Ethics, according to one dictionary, is 'a system of moral principles, by which human actions and proposals may be judged good or bad, or right or wrong' (Macquarie Dictionary).

Traditionally, ethics has been concerned with discovering those principles of behaviour that underlie 'right conduct', or conduct that is morally approved. All actions and policies can be subjected to ethical scrutiny and assessed in terms of their conformity to what may be considered by the assessors to be right conduct in the circumstances. So policies concerned with sustaining biodiversity and the protection of the environment may be assessed in terms of ethical principles and either approved or criticised for the kinds of conduct they imply or for the consequences they are likely to bring. Such an enquiry is a vital precursor to fruitful policy debate because it helps to clarify the nature of moral disagreements as well as possible routes to compromise or more fundamental points of agreement. When the ethical puzzles are clarified, the rest is made more transparent.

Preservation of biodiversity is an issue that arouses deep passions in many people. Such feelings lead to efforts to find and define the moral principles that justify or rationalise those feelings. And, to the philosophical question: '**Why** should species be preserved?' Moral philosophers, biologists and others have attempted to give logically consistent and principled answers. The task of philosophy in general is to discover and define first principles. It is therefore not surprising that the growth of the environmentalist movement over the last twenty years has been matched by a corresponding growth in the philosophical literature concerned with the moral and ethical questions that environmentalism raises.

These sometimes abstract and technical discussions are important for policy because they frequently reveal the unarticulated background presuppositions and premises of those who propose and support particular policy measures. They comprise an arsenal of argumentation behind what might be more baldly stated, in environmentalist literature, as self-evident justifying beliefs. So, in practice, they frequently reappear as moral principles intended to persuade others as to the rightness of a prescription or a point of view.

For example, in the newsletter *Biolinks* issued by the federal Department of the Arts, Sport, the Environment, and Tourism, we are told that: 'Morally, all species and communities have an inherent right to exist' (1991:4). If the Department, and hence the government, takes its own statement seriously this would seem to oblige it to protect species which are 'pests' and which threaten economic activities and human health and comfort. But leaving that aside, an important ethical presupposition about the morally 'inherent rights' of 'all species and communities' is being deployed here to support policies with important implications for all of us.

On the same page of the same issue of *Biolinks,* the argument continues: 'There are two major reasons for conserving biological biodiversity; one moral and the other practical. The practical reason is that the quality of human life depends on biological diversity. From the world's species come our foods, and many of our medicines and industrial products.'

The 'practical reason' for conserving biological diversity in order to serve human interests is sensible enough. But it does introduce the principle, the utilitarian principle, that the practical affairs of mankind and 'quality of life' considerations have a legitimate claim in the determination of conservation policies. Nevertheless, that utilitarian claim has to be set against the 'moral' claims of conservation, and it is in the reconciliation of these two sets of claims that the central issues of environmental ethics are to be found. In what follows, I shall concentrate first of all on the 'moral' claim for conservation of species before dealing with the utilitarian or 'practical' claim.

Although, in its context, it is not offered as part of an explicit ethical-philosophical discussion, an archetypal expression of the moral claim at stake here is to be found in the declaration on preservation of species by the influential biologists and advocates, Paul Ehrlich and Edward Wilson, that '...we and many others think that people have an *absolute moral responsibility* to protect what are our only known living companions in the universe ' (emphasis added) (1991:760). Now this may be interpreted in different ways. It could be taken as saying what most would accept: that we should not deal wantonly or cruelly with other species, especially those capable of experiencing suffering, and we should do all we reasonably can to keep them in existence. The appeal in this is to man's humanity and sympathetic consideration, to the responsibilities that come with man's power over other species and to his duties of stewardship. As such it can be taken quite unexceptionably as a moral recommendation to mankind in general to behave in certain ways towards other species. From that perspective, it is a man-centred or anthropocentric, view in which man is seen as the moral agent; where the moral locus, so to speak, is in man's evaluations, motives and actions.

While this is a possible interpretation of the Ehrlich/Wilson statement, I believe it presents a point of view which goes much further. This is a larger, non-anthropocentric, claim where the focus is upon the *absoluteness* of the moral responsibility which resides *in the nature of things*. It is, in this, similar to the religious duty of obedience to God's will, always and in every circumstances, because God's will is good and this is what God requires of mankind. It is not a *contingent* responsibility that can be set aside when it is incompatible with some vital human interest, but is 'absolute' in the sense that it is simply unavoidable always and in any circumstances.

The idea behind the call for 'absolute moral responsibility to protect other species' is to be found in claims by professional philosophers 'that natural objects and processes are valuable in their own right, in a humanly independent way' (Harlow 1992:27), and that they have 'autonomous intrinsic value' (Rolston 1988:114-117). The common feature here is the idea that other species and the natural world are absolutely or inherently things of value, *irrespective of actual human desires, feelings or attitudes towards them*. This being so, it necessarily follows, so the argument goes, that mankind has an absolute moral duty to cherish that which is intrinsically and autonomously valuable. Before analysing this, it is as well, once again, to distinguish what is being claimed from other interpretations that might be made.

It is quite common, of course, to describe some things as 'intrinsically valuable' for their own sakes, rather than valuable as means to some other end that is (intrinsically) valuable. For example, we might say that being loved by others, or friendship, or the appreciation of art, or the enjoyment of the natural world and other creatures, are intrinsically valuable to us as ends in themselves; whereas the bus or car that takes us to the art gallery or to the wilderness or to a dinner with friends is only ***instrumentally*** valuable. But the point is that, whether of intrinsic or instrumental value, it is *we,* as subjects, who do the valuing. It is the satisfaction of our wants and wishes that constitutes the value. This, the 'subjectivist' view of value, we shall come back to shortly.

But it is not this subjectivist claim to intrinsic value in this sense that is being made in the passages quoted above. It is the quite different claim that other species and natural objects simply *have* 'value' as one of their attributes. This, it is said, is a natural, ***objective*** fact, 'autonomous' and utterly independent of human existence or attitudes; and because they have this *attribute* of value, it follows that certain moral duties are ***necessarily*** laid upon mankind.

CHAPTER 2

Objective or Intrinsic Value

The possibility of an environmental ethic standing independently of human values and human valuing stands or falls on the truth or falsity of the proposition that 'value' is a naturally occurring characteristic of some things or processes. Elizabeth M. Harlow (1992:27) puts it thus:

> There is something both obvious and paradoxical about the central claim of environmental ethics: that natural objects and processes are valuable in their own right, in a humanly independent way. It is obvious because our growing biological and ecological understanding confirms again and again Aristotle's remark that 'Every realm of nature is marvellous'. It is paradoxical because, unlike the 'good' of Aristotle's day, the more modern concept of 'value' is a profoundly human-centred affair – so that the notion of 'humanly independent value' seems perilously close to a contradiction in terms. For this reason, the literature of environmental ethics has been obsessed with the search for a plausible non-anthropocentric 'intrinsic value' in nature; indeed the very possibility of an environmental ethic has been defined in these terms.

In the last sentence of the passage just quoted, Harlow is referring to Tom Regan's remark (1982) that 'the development of what can properly be called an environmental ethic requires that we postulate inherent value in nature'. Consequently, if this postulation is shown to be untenable, the notion of 'inherent natural values', or intrinsic natural values existing independently of human desires and valuations, collapses – and with it the claim that natural things and creatures *necessarily* have equal or privileged moral standing vis-à-vis human interests and human values.

To say that values may exist quite independently of humans is to say that values are particular things or characteristics of things; that they are out there in the world just as objectively as mailboxes or stones, or the *redness* of mailboxes and the *hardness* and *greyness* of stones. And, just as particular red mailboxes or hard, grey stones will continue to exist as such whether or not they are observed or apprehended, so will particular values continue to exist, unaffected by human presence or absence.

THE EMPIRICAL STATUS OF 'INTRINSIC, AUTONOMOUS VALUE'

The problems facing such a theory of independently existing intrinsic value are clearly stated by J. Baird Callicott (1985:259):

> If intrinsic value cannot be logically equated with some objective natural property or set of properties of an entity, independently of any reference to a subjective or conscious preference for that property or set of properties, the only way to rescue the objectivity and independence of intrinsic value is desperately metaphysical: one may say that goodness or intrinsic value is a primitive or irreducible objective *nonnatural* property of some entities. Natural properties of objects may be recognised or discovered empirically or by reasoning based on experience. We know, for example, that an entity is rectangular by immediate experience and that it is radioactive by inference from other immediate experiences (like Geiger counter reports). The nonnatural objective property of inherent goodness or intrinsic value, however, cannot by definition (that is the force of saying it is a nonnatural property) be empirically apprehended or inferred from ordinary sensory experience. It can only be known or discovered, thus, by some mystical intuitive faculty.

This is a question discussed by John Mackie in dealing with the same difficulties of the 'intrinsic value' thesis. Mackie puts it thus: 'If there were objective values, then they would be entities or qualities or relations of a very strange sort, utterly different from anything else in the universe. Correspondingly, if we were aware of them, it would have to be by some special faculty of moral perception or intuition, utterly different from our ordinary ways of knowing everything else' (1977:38).

The 'values objectivist', then, is asking us to believe in a class of things with a form of existence entirely different from the rest of reality, both mental and non-mental. To the question: 'How can we *know* such things?' the objectivist's answer – the only answer he can sensibly give in view of their postulated ontological difference from everything else in the universe – is: 'Not by the ordinary processes of observation and description, but by *intuition*'.

It is clear that the objectivist position is difficult to defend. The truth of objectivist propositions cannot be verified empirically since, in the face of a claim by objectivist X that he has an 'intuition' that Y is objectively (or intrinsically) valuable, or 'a value', the sceptical empiricist has no in-principle way of falsifying the claim; it cannot be tested; it has no descriptive meaning and cannot be verified by observation. But the difficulties do not end there, as we shall see.

As an example of objective values, Mackie offers Plato's 'Form of the Good'. If we know, or intuit, the Form of the Good or the very essence of goodness, our knowledge of that goodness or good thing *impels us* to pursue it for, possessed by that knowledge, we simply can do no other (1977:40):

> An objective good would be sought by anyone who was acquainted with it, not because of any contingent fact that this person, or every person, is so constituted that he desires this end, but just because the end has to-be-pursuedness somehow built into it.

This captures the nature of the philosophical position in the statement by Ehrlich and Wilson that we 'have an absolute moral responsibility to protect what are our only known living companions in the universe'. Protection of our species-companions is an objective good in Plato's sense and, accordingly, 'to-be-pursuedness' (absolute moral responsibility) is built into it.

But we may ask, 'about anything that is supposed to have some objective moral quality, how this is linked with its natural features' (Mackie 1977: 41). So, what then would be the *connection* between the natural fact that other species exist and the claimed moral fact that *not* protecting them is wrong? The one does not logically or semantically entail the other, yet, as Mackie observes, '....it is not merely that the two features occur together'. Thus, in our Ehrlich/Wilson example, the wrongness must in some way be 'consequential'; it is wrong *because* it is failure to protect other species. However, as Mackie says (1977: 41)

> But just what *in the world is* signified by this 'because'? And how do we know the relation that it signifies, if this is something more than such actions being socially condemned, and condemned by us too, perhaps through our having absorbed attitudes from our social environment? It is not even sufficient to postulate a faculty which 'sees' the wrongness: something must be postulated which can see at once the natural features that constitute the [failure to protect], and the wrongness, and the mysterious consequential link between the two. Alternatively, the intuition required might be the perception that wrongness is a higher order property belonging to certain natural properties; but what is this belonging of properties to other properties, and how can we discern it? How much simpler and more comprehensible the situation would be if we could replace the moral property with some sort of subjective response which could be causally related to the detection of the natural features on which

the supposed quality is said to be consequential.

THE SUBJECTIVITY OF VALUES

Or, to paraphrase the last point in the above quotation, how much simpler it would be to affirm the validity of the subjectivist position that there are no values without valuers, which is to say without a personal act or sentiment of valuing; for example, the subjective response experienced by Ehrlich and Wilson and others when they detect 'natural features', say failure to protect species, that offends *their* sense of moral fitness. Values, on such a view, do not consist in entities, qualities or relations of a peculiar kind inaccessible to ordinary ways of observing and knowing, but are simply a species of subjective response to natural events.

We are now in a position finally to explicate the ethical substance of the claim by Ehrlich and Wilson that 'we … have an absolute moral responsibility to protect what are our only known living companions in the universe', and to identify what is there and what is not there. It does not, first of all, point to a prescriptive moral fact or value (a responsibility to protect other species) lodged irremovably, as some kind of independent entity and active agency, in the nature of reality. It does not, in this sense, identify an intrinsic or autonomous value that simply *requires,* by the authority of its mere existence, that we should strive to protect species. Yet it is this conclusion that the word 'absolute' seeks to convey – by suggesting *independence* from human desires and demands. When, as we have seen, the claim to independent fact falls by the wayside, what we are left with is a statement that reveals two things: first, that it is a straightforward human demand that we should all act to protect other species, and which illegitimately recruits an unsustainable metaphysical device in order to persuade; and, second, that its authors are telling us, implicitly, that they hold certain values – the preservation of species – which animate their policies and actions and which, when realised, elicit in them desirable subjective responses.

How such subjective responses come to be culturally and historically 'conditioned' and their regularities expressed in individuals and communities as 'moralities' and 'ways of life', is a separate question. But it is a question amenable to ordinary investigation and hypothesis and does not require the postulation of different orders of reality and special or mystical ways of knowing or 'intuiting' them.

Our daily moral judgments and expectations – that 'one should not steal'; that 'neglecting children is irresponsible'; that 'cruelty to animals is bad'; that 'political liberty is good'; that 'species should be preserved,' and so on – although frequently offered in a verbal form that is 'objective' and prescriptive – can nevertheless be unmasked as philosophical and rhetorical short-cuts that conceal their social-moral embeddedness and natural history. The sort of dissection sketched above reveals, in short, their connection to culturally contingent desires, and demands commandeering a mantle of 'objective' authority that is, in fact, plainly human and social in origin and not extra-human. The heart of the matter, then, is the bogus claim for a form of moral authority lodged in nature itself and beyond the challenge of humanity and its competing social and moral interests. Mackie summarises this well:

> We get the notion of something's being objectively good, or having intrinsic value, by reversing the direction of dependence here, by making the desire depend upon the goodness, instead of the goodness on the desire (1977:43).

I might add that the attempt to find moral prescriptiveness and ethical imperatives *in the nature of the world* is frequently a political device intended to sacralise, or to conceal, or to place above scrutiny and criticism, essentially human interests and policies (a device finding its classic expression in Plato's *Republic,* where objective goodness is discovered in the political arrangements Plato happens to prefer).

The important point, therefore, is that when the metaphysical underpinnings of 'inherent right to exist', 'absolute moral responsibility to protect', and 'intrinsic, autonomous value' collapse, what we are left with are nothing more, and nothing less, than historically-conditioned human desires and subjective values, none of which can claim a transcendent status and all of which must struggle on a common level for achievement or realisation.

Values in general, then, are subjective responses to those things or events which are esteemed by us because they are useful or satisfy various desires. 'Moral' values refer to forms of conduct that we esteem and (usually) prescribe.

In concluding thus, I would therefore agree with Carruthers that it is false to claim that 'moral values form part of the fabric of the world independent of our minds' (1992:195).

CHAPTER 3

Moral Values & Human Interests

Moral values deal with questions of conduct in conformity with duties, rights and responsibilities. We know we are in the realm of moral values when conduct entailing duties, rights and responsibilities invites assessment in terms of approval or disapproval. Those forms of behaviour that don't entail duties, rights and responsibilities may nevertheless invite approval or disapproval on other grounds and in relation to non-moral values. Playing a game well may invite approval from those who value good game-playing, but that is not a moral value.

Moral values come into play when human beings interact with each other, as individuals or groups, in ways that affect their respective interests. So, in my dealing with you, you may come to the conclusion that I have acted 'meanly'; whereas I might judge you as having acted 'generously', and each of us is giving a moral characterisation of the other's conduct.

Thus, if a particular tree has no value for anybody, or if its presence or absence is of no interest to anybody, my behaviour in cutting it down, or not cutting it down, is morally irrelevant. The tree itself is not the possessor of a moral right either to be preserved or to be cut (Carruthers 1992:1). On the other hand, if the tree is on somebody's property, and therefore part of a system of rights held by the property-owner and valued by him, the question of my cutting it down is a moral issue, not simply because the tree is valued, but because my conduct in cutting it down would have infringed rights. It is the landowner who has moral standing because he is a rights-bearer, and I should not have cut ***his*** tree down. The tree does not have moral standing. The tree ***matters*** only in an indirect way because it signifies part of the substance of the rights concerned. It is the landholder's rights and interests and the duties towards the landholder that they imply for me which have the moral standing and which therefore raise issues of *moral* values and proper or improper conduct.

But we can also say that, in the absence of formal rights, there may still be interests of a moral kind which it is possible to infringe. Suppose the tree we have just been talking about is a particularly handsome one that nobody owns but which gives many people great pleasure simply to observe and enjoy. Morally, the tree has no significance in itself. However, if, knowing that that tree is valued by many people in this way, a person wantonly destroys it, can that person's action be described as reprehensible and morally wrong? Most would agree that it would be wrong if they subscribed to the moral principle (i.e. had the moral value) that conduct should take account of the interests of others. By destroying the tree, that person has injured the interests of others. Thus the tree itself is only indirectly relevant and the real focus of moral concern is not ***its*** destruction but the foreseeable *consequences* of its destruction for other interests and the conduct of the destroyer in bringing about those consequences.

The general principles are, then, that:

(i) Moral values are that sub-class of subjective responses to forms of conduct that conform to or diverge from moral principles or beliefs that we adhere to.

(ii) Natural objects or processes cannot have moral status in their own right, but moral duties may be indirectly involved when objects valued by persons may be damaged by human conduct. ***A moral issue arises when conduct has consequences for human rights and interests.***

MORAL BELIEFS AND MORAL JUDGMENTS

Moral values are essentially subjective responses or attitudes, focused on human conduct. So an action or an item of behaviour, considered simply as such, does not have moral value as one of its attributes. The occurrences of moral values or moral evaluations take place in the minds of those who observe and respond to actions, and then only if those actions, the reasons for them, and their consequences, can be contextualised and assessed in terms of their conformity or otherwise to principles of conduct ***valued*** or adhered to by the observer. It is frequently the case that actions or behaviour cannot be morally evaluated until we know more than the acts themselves; that is, until they have been 'placed' or contextualised. So an act of cutting down a tree may be totally unblameworthy if it is done by a forester authorised to do so; but exactly the same action will be assessed as blameworthy, in the circumstances, if it is an unauthorised act of 'vandalism' on private property. Killing a man, if only the mere action itself is considered, cannot be morally assessed until we know whether it was in legitimate self-defence or a plain act of murder. Moral evaluation takes place when the conduct, and the motivation of the conduct, have been tied together.

CHAPTER 4

Moral Values & Animal Rights

In this chapter, the problem of our moral responsibilities to other species, particularly the higher animals, is tackled. Part of the problem is to reach a consistent point of view on the question of animal rights and to integrate the conclusions with the principles governing human moral action in general. But I must first explain my view of 'rights', including the idea of 'natural' rights.

RIGHTS AND 'NATURAL' RIGHTS

'Rights', in general, are nothing more than claims by individuals and groups that can be made good within a moral/social order capable of prescribing and enforcing them. A right asserted but incapable of being acknowledged or made good is not a right at all; it remains a mere demand or assertion. Animals are incapable of formulating or prescribing rights or of making claims under them and enforcing them. Only human social/moral orders are so capable, and although animals may be the operational 'beneficiaries' when certain claims are made on their behalf and enforced, as they can only be enforced, by human action (e.g. prevention of animal suffering), this is not proof of ***their*** rights, but of ours.

The definition of rights offered above rules out any possibility of 'natural rights', in the sense of claims which, ***in the nature of things***, are always successfully asserted and made good – claims which are literally and actually undeniable. Such a position is another form of the 'absolute moral responsibility', or 'inherent right to exist' arguments that we have already disposed of. 'Natural rights', 'absolute moral responsibilities' and 'inherent rights to exist' are devices which attempt to conceal the unavoidably contingent or conditional nature of all rights and responsibilities; conditional, that is, on the real world likelihood or otherwise of an actual claim or assertion of a right being made good or enforced by human agency within a moral community. Animals, in common with human beings, have no 'natural' rights; and insofar as their existence or welfare may be at stake in their interactions with mankind, they have no inherent claims, no protections and no rights except those constraints on human behaviour which mankind erects to serve *human* interests and *human* values.

PRINCIPLES OF CONDUCT, 'TASTES', AND MORAL ORDERS

In arguing for the subjective nature of values and evaluations, does it follow that one set of values is just as good, just as acceptable, just as workable, as any other? I have argued that 'value' is not a quality or attribute that can be discovered in the external world but a subjective response to events or situations that satisfy our needs or desires, including our moral needs and desires. Is it, then, simply a question of 'taste'? As Carruthers puts it: 'This is the question whether moral judgements are merely subjective expressions of attitude ... For example, if making the statement hurting animals is wrong' is a bit like saying "I dislike cheese", then there is really nothing more to be said' (Carruthers 1992:4).

But there is an important difference between matters of taste and matters of conduct in accordance with beliefs or principles. In finding our way through the world and through life with others it is vital to our interests that the world makes sense, that the conduct of others be broadly predictable, and that our own conduct should have the kind of coherence that comes from adherence to beliefs, and rules of conduct effecting those beliefs, that hang together in a more or less orderly and consistent way. Inconsistency and contradiction within ourselves is psychologically painful to us, it sets up 'cognitive dissonance', whereas we may have no difficulty at all in accepting our liking for milk and our dislike for cheese, even though both are dairy products.

Insofar, then, as moral values tend to coalesce in systems of principles or beliefs – in ethical systems – that have a strain towards internal consistency, there is scope for those who share certain fundamental moral principles to examine their differences about particular issues of values in terms of those fundamental principles. This, indeed, is characteristic of rational argument and democratic discussion. If we agree, for example, that species should be preserved unless some vital human interest is at stake, we can rationally discuss what constitutes 'vital human interests' and, given some measure of agreement on this, go on to refine our views and rationally discuss how specific situations are to be defined and dealt with.

But to say this is not to accept a doctrine of objective moral values somehow existing independently of humanity and waiting to be discovered. It is rather to accept their inevitable subjectivity but (subjectively) to value consistency between moral beliefs. Also, to recognise this does not necessarily mean that they are unchangeable and beyond reconciliation when people disagree initially. The human strain towards consistency and rationality and the avoidance of cognitive dissonance, open up opportunities for fruitful discourse, the rational comparison of moral stances, and the possibility of resolving inconsistencies and conflicts:

> We should perhaps agree, indeed, that our moral beliefs can only really be acceptable if they form part of a coherent body of such beliefs, linked together by general principles having at least a powerful intuitive appeal. It follows that a considerable part of our task, when it comes to determining the appropriate moral treatment of animals, will consist in seeing how principles concerning such treatment might fit acceptably in to an overall moral theory (Carruthers 1992:5-6).

MORAL ORDERS AND MORAL VALUES

We referred earlier to the connections between moral values and the rule-governed behaviour that gives effect to moral values in shaping what we then call 'conduct'. A morality is characterised by the nature of the rules that control the ways in which people conduct themselves in dealings with each other. We speak, for example, of Christian ethics, or Muslim ethics, or business ethics, or professional ethics, and we make those distinctions in terms of the characteristic moral rules that are acted out in the interactions of the members of those communities, or between them and the members of other communities.

Thus, to speak of an actual morality or system of ethics is to refer simultaneously to the community (businessmen, physicians, Christians, Muslims, etc.) in which that morality, that set of moral values, is acted out. A degree of reciprocity, common observance and the capacity for moral agency (i.e. conduct consciously or habitually conforming to moral rules) are of the essence of a moral community – as distinct from a herd – and essential to its existence. Otherwise we are simply confronted with instinctive or reflexive behaviour rather than conduct governed by moral rules emerging from a community and constituting an element of its culture capable of being learned and handed on.

But behaviour or, more accurately, conduct that is the consequence of apprehending both the generality and the specificity of rules and then shaping action to conform to them in infinitely variable situations, requires a capacity for abstraction, matching, evaluation, anticipation and future-imagining that is pre-eminently conscious – at least until rule-absorption-and-following has become habitual – and rational. It entails mental faculties of a high order.

So a certain level of mental capacity, and rationality exercised by a deliberating participant within a moral order or community, are the essential conditions of moral performance, and minds so capable and so located are the only locus of moral values.

THE MORAL STANDING OF ANIMALS

We concluded above that objects (e.g. trees, houses) cannot have moral status in their own right but they may acquire indirect moral significance when involved in actions affecting human rights and interests. Clearly, in at least this limited sense, the same would be true of animals. If, for example, somebody rides my horse without my permission, the rider morally transgresses *me*, but not the horse. The horse is only ***indirectly*** involved in the wrong act; I am the one whose rights and interests have been damaged.

But most would agree that the horse would figure ***directly*** in a moral issue if the rider should treat the horse cruelly while riding it. Could we not say, quite straightforwardly, that the horse is a victim of a moral outrage and that the rider has acted wrongly towards it? It has been the settled view for some time that this would be so.

But what if the rider had borrowed my horse in an emergency to ride for help for his child who had been seriously injured in an accident, and that he had had to thrash the horse mercilessly in order to get aid as quickly as possible? Does this make a difference? Can we then accept that the cruelty is justified and therefore not wrong; that the child's need overrides the horse's interests or 'rights'? We might even conclude, despite the fact that the horse had suffered grievously, that it had not been treated ***cruelly*** because the rider's motive was not to hurt the horse but to get help quickly.

Nevertheless, in other circumstances, it may make sense to us to speak of cruelty to an animal, but it doesn't make sense to speak of cruelty to a house or a tree (although the idea of 'wanton destruction' of objects, and 'vandalism', does make sense; but that is a separate issue). As utilitarian philosophers have argued, it is the capacity for suffering of animals which makes the difference and which awakens our moral sensibilities and sympathies when they are hurt. So, gratuitously to cause animal suffering is now overwhelmingly regarded as wrong, and the higher up the evolutionary scale the animal, the closer its resemblances to mankind, the greater the condemnation when it is made to suffer. But what we mean by 'gratuitously' in this context is that the suffering ***is not in any worthwhile cause***; it serves no purpose, or no good purpose. Whereas if the purpose is a good one (e.g. to save an injured child's life), most would agree that causing the horse to suffer is morally justified. In other words, common sense morality accords some moral standing to animals but it is not equivalent to the moral standing of human beings. The common view is that, in the last resort, non-trivial human interests must prevail over animal interests.

The fact remains that we regard unjustified cruelty to at least some animals as wrong, and this fact must be accommodated in any coherent moral theory. Also, we said above that cruelty to the horse involves the horse ***directly*** in a moral issue. The question to be answered is whether this is sufficient to give the horse itself moral standing, or whether more is required; and, if so, whether it can be delivered.

Peter Singer (1975; 1979) and Tom Regan (1984), claim that the moral standing of animals should be affirmed. Singer, for example, bases his affirmation on the contention that a coherent ethics must be based on the principle of equal consideration of interests. If it cannot be shown that an individual or a species differs ***in some morally significant way***, then those individuals and species have moral standing and their interests are entitled to equal consideration. The characteristics of animals that differentiate them from humans – four leggedness, fur, lower intelligence, inability to speak, etc. – are morally irrelevant. Just as we now treat the colour of a man's skin, or differences in intelligence between people, or deformed human bodies, as morally irrelevant, so should we treat animal differences as morally irrelevant. For, in respect to the characteristics that are morally relevant – sentience, and capacity for pleasure and suffering – animals are our equals and therefore entitled to have their interests considered equally with our own within our moral order.

How are we to reconcile this with the common sense moral view, in our example above, that it is permissible to ride a horse cruelly if it is necessary in order to save the life of an injured child? And must we not also give some weight to the unique human capacity for imagining or anticipating the possibility or certainty of pain or humiliation; for suffering the anxiety that comes from knowledge of impending misfortunes and unpleasantness? It seems that human suffering can be larger in scope, more exquisitely varied and deeper in quality than animal suffering, and that it has an extra psychic dimension that is not present in animals.

Singer acknowledges and tries to accommodate this by incorporating a view of 'higher' and 'lower' forms of life that would justify our preferring to save the life of a human being rather than an animal. But such a device clearly betrays essentially human preferences and evaluations in defining what constitutes 'higher' and 'lower'. It smuggles questions of value and degrees of estimation into an issue that should be determined, if his basic argument is sound, solely in terms of the objective presence or absence of particular characteristics. Either animals are entitled to equal moral standing because they are sentient and can feel pleasure and pain or this is not sufficient and something else, which may or may not be present, is required. 'Degrees' of moral standing in terms of 'higher' and 'lower' forms of life are a distraction from that main issue.

So, if the question of 'higher' and 'lower' forms of life is irrelevant here, the fact remains that Singer's perspective drives him relentlessly towards ignoring common sense and to affirming the equal moral standing of animals. It follows that there is no moral basis for choosing to save a threatened human life rather than a threatened animal life. This would require us to regard the slaughter of animals for food as morally equivalent to slaughtering humans for food.

It seems, then, that Singer and Regan and others who argue for the equivalent moral standing of animals have an insuperable problem, at least of persuasion if not of abstract theory. But even in theoretical terms there are other grave problems, as we shall see. Nevertheless, even if one disagrees with their conclusions, we should give reasons why we think we are obliged to avoid causing animals unnecessary suffering. I will return to that question when we have dealt with the issue of the 'moral equivalence' of animals other than man, and that requires some discussion of what it means to be a moral actor within a community.

MEMBERSHIP OF MORAL COMMUNITIES

The phrase 'the moral standing of animals' suggests an ambiguity – it may refer either to status as fully franchised members of a moral community, or to the standing that the members of a moral community accord to those who are less than full members. In the first case the claim is to full and equal membership because all of the basic qualifications for membership have been met, and in the second case it is standing of a quite different kind which consists in the obligations that the members of the moral community place upon themselves in their dealings with non-members. So, to settle the ambiguity, we must begin by noting the conditions that must be met for full membership of a moral community.

Any reasonably complex, pluralistic society is a mosaic of communities and moralities, often inter-penetrating, and held together by observance of some fundamental, overarching rules or laws. We must be careful not to be unduly solidarist in attributing strict observance of moral rules to societies or to communities (regional, ethnic, occupational, religious, etc.) within societies. Nevertheless, a morality, or the moral order that obtains within a community, is *sui generis*. It is the system of rules of conduct that constitutes it *as* a community. It has emerged from their living, working, praying, etc., together, and, in a sense, *is* their living together. This is not to say that all its members necessarily abide perfectly by its rules, but there must be substantial conformity to some core rules, or its communal features and the conforming conduct that sustains its morality and identity will disappear. Nevertheless, to have standing as a member of a moral community one must be capable of apprehending the rules and either enacting them or breaking them by conduct constituting moral performance and entailing rational agency. It will be recalled that we concluded, in our discussion of principles of conduct and moral values, that *'a certain level of mental capacity, and rationality exercised by a deliberating participant within a moral order or community, are the essential conditions of moral performance, and minds so capable and so located are the only locus of moral values.'*

The question, therefore, is whether animals can meet these conditions. If they cannot achieve, or have the potential to achieve, moral performance they cannot be paid up members of a moral community and entitled to equal treatment and equal rights. So, do animals have moral values and are they rational agents capable of moral performance?

ANIMAL INTELLIGENCE AND MORAL AGENCY

The capacity of animals, from pigeons to gorillas, to learn and reproduce complex patterns of behaviour involving responses to symbols, to fashion and use tools to get food or to lure victims, to cooperate in hunting, to hide food and defer gratification in order to 'deceive' other members of their species about its whereabouts, are all well attested in the zoological and ethological literature. Chimpanzees may break and make sticks to size and shape in order to dig termites from a mound, finches may use twigs to prise grubs out of holes in trees, and sea otters may use rocks to break open shell-fish.

We can only speculate about the inner, mental life of animals other than man. We do not know whether primates, for example, can imagine alternative futures or foresee the consequences of possible actions and formulate plans and actions accordingly. Many of those who reflect on animal behaviour are sceptical. Much, if not all, of what is observed in animal behaviour may be explained in terms of natural selection of genetically programmed response patterns and learning. Nevertheless, final judgments about their thinking and their degree of self-consciousness must be withheld.

As for the emotional life of animals, there seems to be no doubt that we can identify specific emotions. We have already referred to their capacity for suffering. Dogs clearly 'enjoy' the return of their masters and mistresses after an absence. Chimpanzees seem to enjoy mutual grooming and to 'prefer' bananas to carrots. Animals, then, reveal a range of emotions, desires, appetites and preferences. Even insects seem to have their 'tastes' and preferences. Can we infer from this that animals – chimpanzees for example – have 'values'? Is apparent enjoyment or preference indicative of the existence of a system of values? Here we hesitate. Except in a loose way, we tend not to say that chimpanzees 'value' bananas or grooming, or that mosquitoes 'value' blood, because the word suggests a degree of self-conscious mentality, active hierarchies of estimation and deliberative appraisal and weighing of attributes that may be more parsimoniously explained in terms of programmed and reflexive behaviour or 'instincts'. But deep scepticism about animal valuing, in a sense comparable to human valuing, is not proof. It may be that animals do value things and processes. However, we can leave that question open and go on to examine the more important question whether animals, including chimpanzees, can have ***moral*** values and (crucially) *culturally* hand them on and reproduce them in later generations.

Here, we can be more confident in denying that moral values form part of the fabric of non-human minds, including chimpanzees' minds. The latter can be enraged by the behaviour of other chimpanzees. But only some forms of behaviour are 'conduct'. Conduct is conformist or deviant deportment in the light of a system of rules of behaviour prescribed and handed on by the participants in a social or moral order continuing over several generations. Chimpanzees may observe and respond, favourably or unfavourably, to the behaviour of other chimpanzees or man; but they do not 'judge' it, as conduct is judged, in terms of its conformity to abstract rules. And yet this is the essence of moral evaluation and in its absence there can be no acceptable sense in which it can be said that animals have moral values.

It is essential for moral performance, as we have stressed, that moral agents should be capable of apprehending the generality and specificity of rules of conduct. They should be able deliberately either to conform to or deviate from the rules in shaping their conduct. They should also be able consciously to reciprocate in rights-acknowledging exchanges with others identified as rights-bearers. It is the rational capacities underlying this kind of adaptability and deliberative action which are absolutely crucial for moral performance, and the evidence shows that those capacities are not present in animals, not even the higher primates. No matter how complex and flexible the social behaviour of chimpanzees, for example, there is no evidence whatsoever that their behaviour reveals the apprehension and acting out of general rules of conduct ***established*** by a chimpanzee community and incorporated by a process of instructive socialisation in the behaviour of its members. Nor do we find, as with human communities, the wide variability in conduct that issues from different moral traditions. Although there is some evidence for the learning and handing down of discoveries in the use of tools from one generation to the next, and evidence for the importance of models in the learning of effective maternal behaviour amongst the higher primates, there is no evidence, and no mechanism available in the absence of language and oral traditions, for the handing on of general rules of conduct from one primate generation to the next, or from one primate group to another. The presence of a many-sided culture among mankind, of which moral values and traditions are of supreme importance, and its absence from the life of even the most intelligent animals, are critical.

We must conclude, therefore, that animals other than man do differ from man in morally significant ways, and to a huge degree. Animals other than man are incapable of forming or having moral values and of exercising moral judgment in the senses that we have seen to be required. They are incapable of the moral conduct and rule acknowledging performance, including reciprocity, that are necessary for membership of a moral community. They must therefore be denied the moral equivalence and common rights that are the mark and consequence of such membership; and denied, also, the moral standing that is claimed for them by writers such as Singer and Regan. But it is nevertheless consistent with this conclusion for mankind to accord to animals the restricted form of moral standing that accrues when mankind places constraints on its own conduct in its dealings with animals. I will have more to say on this in a moment, after discussing a parenthetical difficulty arising from the foregoing argument about the connections between rational moral agency, moral performance and membership of a moral community.

This is the question whether incapable humans should enjoy the full rights of membership of human moral communities.

If rationality and rule-apprehension-and-following are beyond the capacities of a new-born baby, or a mentally handicapped or comatose person, this seems to mean that they are incapable of moral performance as I have defined it and should therefore be ineligible for membership of the moral community. In practice, of course, we do not deny their membership status. Why not?

An answer offered by Carruthers (1992:110-121), adapting an argument by John Rawls, is that since every human being is the child of a human being and rational agent, all human beings should have the same rights and be accorded full membership of the moral community. But since no animals are descended from rational agents, they cannot enjoy the same rights and membership.

A separate argument depends upon the clear differences between human beings and other animals, and the slippery slope that awaits us, in practical terms, if we begin to qualify human rights and moral standing according to human deficiencies and handicaps; if we sought, that is, to measure out reduced portions of human rights in proportion to degree of human rationality. As Carruthers puts it: 'For to think and speak in terms that withhold moral rights from some human beings is to invite people to try to draw yet further distinctions – for example, withholding rights from those who are sexually or intellectually deviant or from those whose intelligence is low.' And later: 'The only way of framing rules that we can live with, then, is to accord all human beings the same basic rights – that is to say, moral standing' (1992:118).

ANIMAL RIGHTS

In denying the moral equivalence of animals and the possibility of their membership of the moral community, we nevertheless agreed that mankind can, and does, acknowledge that it is wrong to treat animals badly or cruelly except when it may be justified by some significant human benefit. And we noted, also, that even Singer and Regan accept this. Animals, at least the higher animals, are clearly capable of suffering and it is now commonly regarded as morally repugnant for mankind to cause suffering unnecessarily or gratuitously. Accordingly, civilised societies now have laws designed to protect animals and we must seek an explanation of this within our theorising. Carruthers suggests two main reasons. The first is that we place restrictions on our treatment of animals because many, if not all, persons are distressed by animal suffering and it is wrong for others to cause them distress by treating animals badly:

> Since many people have concern for animals, and are deeply distressed at seeing an animal suffer, this may place on us an obligation not to cause suffering to animals, except for powerful reasons. This would not be because needlessly causing such suffering would violate the rights of the animal, any more than someone who defaces a building violates the rights of the building. On this approach animals, like buildings, would have no direct rights or moral standing. Rather, causing suffering to an animal would violate the right of animal lovers to have their concerns respected and taken seriously (1992:106-7).

The second reason he offers is that cruelty to an animal and indifference to its suffering are indicative of defects in human character, and we therefore punish the infliction of suffering and seek to eliminate or control the degradation of human character that accompanies it. We value those people who, in general, are capable of sympathetic and benign responses, such as responding instantly and kindly to the distress of an animal. Such a trait is valuable in human society and its opposite is something to view with apprehension because of the kind of character it suggests. Thus, 'the moral value of the virtue, in so far as it manifests itself in our treatment of animals, derives from its connection with our treatment of human beings' (1992:154).

On this view, then, 'animal rights' are not claims that may be made by animals in their own right. Animals have no entitlements by virtue of their being animals, as human beings have by virtue of their being human. In this sense, animals have no moral standing as the members of a moral community may have moral standing. But they do have standing in the strictly limited and qualified sense that, as a matter of fact, they are accorded special moral treatment by virtue of their capacity for suffering, and this issues in the two main kinds of duties towards them (discussed above) that human societies demand of persons, ***in the service of human morality and human interests.***

I believe that the foregoing arguments, if one accepts the impossibility of 'objective values' and the non-existence of moral values among animals, imply the untenability of a non-anthropocentric ethics because the only remaining subject-matter of ethics is human conduct and the struggles between men to establish particular moralities or rules of conduct. I would therefore endorse Cora Diamond's remark that 'our *hearing* the moral appeal of an animal is our hearing it speak – as it were – the language of our fellow human beings' (1978:478). When we cease to hear the [moral] language and appeal of our fellow human beings, animals will be among the first to suffer. So, somewhat paradoxically, both animal and human welfare depend upon the denial of the supposed 'moral equivalence' of other species and mankind which is frequently used to justify the preservation of species, irrespective of the effects upon human interests.

One of the disturbing trends of that line of argument is its use to parody legitimate human concerns by painting them as entailing the suppression of 'animal rights' and as therefore wrong on *that* ground, whereas the real ground of disagreement is between two human value positions. In the hands of some, it leads to the view that the human species is a pest which deserves to be eliminated (with a few special exceptions, of course). Thus we find research biologist David Graber reported as saying of mankind: 'We have become a plague upon ourselves and upon the Earth... . Until such time as Homo Sapiens should decide to rejoin nature, some of us can only hope for the right virus to come along' (Postrel 1990:28).

Our treatment of animals cannot be unrestricted, casual and indiscriminate, for this would be an offence against the human moral community. But it does not follow from this that animals have moral standing, except in the special sense detailed above. Insofar, then, as we may speak in casual conversation of 'animal rights', this refers to nothing more and nothing less than the range of liberties and restrictions that mankind allots to ***itself*** in its dealings with animals. There is nothing morally immutable or guaranteed, therefore, so far as animal 'rights' are concerned, and no undeniable claim on behalf of their welfare which cannot be overturned by a countervailing claim in favour of human welfare.

USING ANIMALS AND KILLING ANIMALS

The ethical significance of our treatment of animals, it has been argued, consists in its possible effects on human character and what it reveals about human character. Animals should be saved from unnecessary suffering because to act otherwise is to be cruel or cruelly indifferent, and cruelty is an unacceptable character trait. Suffering, and causing suffering, should evoke in us feelings of anguish and remorse; and if they don't, we are morally deficient. What we hope of our society is that it will inculcate, to the widest possible degree, immediate indignation and revulsion against those who are cruel and wanton about animal suffering; and that it will inculcate and encourage sympathetic treatment of animals.

But when, if ever, is animal suffering necessary or acceptable? When it is unavoidable in meeting a significant human need – we gave the example earlier of forcing a horse to exhaustion in order to get help for a seriously injured child. But what are the criteria of 'significant human need' and how are they to be justified? Such questions are unanswerable abstractly and absolutely. There is no *a priori* answer, but only 'case law' and precedents; the slow evolution of custom and convention in such things, that might or might not reveal implicit and discovered principles about what constitutes the kind of conduct that is acceptable in dealing with animals and that may sometimes be enshrined in formal law. But always recognising the principle that animals do not have moral standing in their own right and that ***the only measure of appropriate treatment is what is good for human society and human character.***

Causing the death of an animal is likewise acceptable, provided similar guidelines are observed; which is to say, that the killing should not be wanton or wasteful and should, as far as possible, avoid causing suffering. Again, the criterion is the implications for human motives and character. In caring about the killing and the manner of killing the focus of our moral concern is what it means for the character and well-being of our society and the people in it.

Obviously, there will always be borderline cases where it will be difficult to decide whether a particular practice – battery farming of hens, hunting for sport, experimenting on laboratory animals – is reprehensible in these terms or not. Probably, in these examples, the human participants and the history of the practice reveal no failure of character, and generate no defect of character, except for the occasional individual who may behave badly – but that is possible in any field of interaction, whether with humans or animals. Also, of course, there may be huge differences in the way different cultures deal with animals, and we are not necessarily entitled to draw conclusions from such treatment – using dogs for food, for example – that the humans concerned are likely to lack decent feeling for other people or to be sadistic towards the animals.

For some people, preservation of species and individual members of species are moral absolutes, which is to say that they believe that this must be one of the objectives of their conduct. Killing or using animals is always to be avoided and extreme measures to protect and preserve are obligatory, even at risk of significant human interests and human life. This is a stance that goes beyond ordinary, or even severe, strictures upon prevention of cruelty to animals, and the discussion above is intended to sketch the ethical approach of the writer, especially in situations where causing the death of animals may be necessary to save aspects of the environment or the survival of animal species. As we shall see later, the major threat to the survival of small native animals in Australia is the destruction being wrought by foxes and cats, and it may well be that the survival of such native species depends upon exterminating their predators.

CHAPTER 5

Moral Subjectivism & Moral Principles

It is sometimes assumed that the 'values subjectivism' that has been expounded here leads, inevitably, to nihilism and moral scepticism. But that does not follow at all, if it is being claimed that, as a matter of ***causality***, to believe in the subjectivity of moral values necessarily determines what those values will be and how we will conduct ourselves; or that such a believer is condemned to a life of moral indeterminacy and incoherence. A belief in the subjectivity of values does not support a particular view of what our values ***should*** be or must be. Nor does it mean that one can simply pick and choose, as if one were an ahistorical and unconditioned free agent, any morality at all or no morality at all.

Mankind cannot help being a species that generates moral rules and moral orders, because mankind is a social species and is probably programmed for responses, such as sympathy, socially-judgmental categories of 'fairness', and perhaps a range of 'tribal' and familial responses, upon which moral rules come to be erected (Wilson 1993). Otherwise there could be no human society and thus no possibility of maintaining the conditions necessary for individual and species survival.

But dispositions of the human species to respond with sympathy or affiliative impulses, do not, in themselves, determine how a particular (already-existing) culture will mobilise and order those individual dispositions into specific forms of conduct. In other words, a species containing those dispositions presupposes a social-cultural milieu and a pre-existing moral order capable of evolving and reproducing them through traditions, customs, instruction and learning. And, since different kinds of moral orders and cultures express and shape these dispositions in different ways, it is obvious that morality cannot be reduced to psychology or biology, or be seen to be, in its origins, a rational construction of human reason; although human reason may be able to discern and articulate the general rules implicit in a 'natural' (i.e. existing) morality; just as we may be able, by reason, to discover and express the grammatical rules that give structure and predictability to a natural language. But people do not need to first analyse the structure of their language in order to use it correctly. As with morals, language is not the product of reason or design but a spontaneous system; a system, in the case of language, of (semiotic) conventions arising from countless communal encounters and the imperatives of needing to communicate in order to live together and to exploit the advantages of doing so. Adam Smith captures the point thus:

> But though reason is undoubtedly the source of the general rules of morality, and of all the moral judgments which we form by means of them; it is altogether absurd and unintelligible to suppose that the first perceptions of right and wrong can be derived from reason, even in those particular cases upon the experience of which the general rules are formed. These first perceptions, as well as all other experiments upon which any general rules are founded, cannot be the object of reason, but of immediate sense and feeling. It is by finding in a vast variety of instances that one tenor of conduct constantly pleases in a certain manner, and that another as constantly displeases the mind, that we form the general rules of morality (1982:320).

Just as there can be no 'justification' for the existence of a particular species and its modes of behaviour, neither can there be a 'justification' for forms of moral conduct over and above the fact that they occur. Justification, if it is to be spoken of at all, is no more than conformity to the prevailing moral rules; it is not measuring actions against some standard, which is not itself a rule, that exists independently of the rules. To repeat, there are no values inhering in things or processes and no moral prescriptions or imperatives of any kind to be discovered in them. As Peter Singer has put it (1993:188):

> If the universe has not been constructed in accordance with any plan, it has no meaning to be discovered. There is no value inherent in it, independently of the existence of sentient beings who prefer some states of affairs to others. Ethics is no part of the structure of the universe in the way that atoms are.

So, one can believe in the subjectivity of values and nevertheless, without contradiction, subscribe to, and act in accordance with, a highly restrictive system of moral values and religious principles, on the one hand, or principles of a distinctly libertine kind, on the other. The only scientific questions that arise are how we come to develop particular moral values or preferences, to adopt particular codes of ethics, and how we handle the consequences. Those are psychological, sociological and historical issues, and valid answers will not prove that one way of life should be pursued rather than another. Nevertheless, the fact remains that we make moral judgments and try to enforce moral rules in accordance with the system of moral values to which we adhere, and we cannot give an 'ultimate' justification for the making or enforcement of any of those judgments or rules. It is therefore idle to speak of the ***validity***, in the sense of the truth, of a moral prescription or the rightness of an action, except insofar as the latter simply means its conformity to an existing and accepted rule.

Nevertheless, belief in the subjectivity of moral values does not mean that one is committed to moral disorder or values incoherence. If one believes in the desirability of the coherence of moral principles, or, as some moral philosophers put it, 'reflective equilibrium', one is not at liberty to be capricious. As we have said above, an ethical system that deserves to be called such is not simply a random collection of varying moral 'tastes'. It articulates itself as moral principles having a greater or lesser degree of internal consistency. Moreover, if there is to be any moral discourse at all, it must begin with an assertion of some principle or other and proceed to test that principle against consistency with other principles and with actual conduct. Such a procedure leads very swiftly to qualifications, according to changing circumstances, of the application of principles to real cases. For example, we might want to qualify the moral principle that promises should always be kept by excepting those cases where promises have been extracted under threat. The principle and its power remain, but it is no longer absolute.

It needs also to be said that we are necessarily born into, and raised in, an established and ongoing moral order from which we must draw the only moral nourishment that is available to us and which must therefore provide our only repertoire of moral rules. So our moral thinking can never be wholly free or uncharged with the moral atmosphere that surrounds us. But even the most limited, cohesive and isolated of societies contain pockets of incoherence and unsettled moral dilemmas which provide the seeds for moral argument and disagreement, carried on, unavoidably, in terms of the moral principles that have currency there. But argument and real world experience, or interaction with other societies and other moralities, expose those principles to change and modification. We will later be concerned with examining the institutions of our own society that permit that principled argument to be carried on in relation to the moral issues that surround the preservation of biodiversity.

SUMMARY

I have argued that 'value' is not an objective, autonomous or intrinsic attribute or characteristic of any object or process. When we speak of 'our values', or what we find 'valuable', we are, in fact, pointing to a relationship between our subjective experiences and other things or processes. So, to have a value, or to value something is to refer to a subjective feeling of pleasure or satisfaction when we possess or experience that which is desired or needed.

Along the same lines, moral values are a sub-class of values which refer to the subjective experience of satisfaction or pleasure we feel when contemplating conduct (by others or ourselves) which conforms to rules of conduct to which we subscribe or which we approve.

Values of all kinds, then, are to be found not in particular attributes of things in the world, but in our subjective attitudes and reactions to things or processes in the world.

The rules of conduct which express moral values issue from our living together. It is these rules and the social forces behind them that make moral demands of us. It is our commitment to the rules, and not other facts or events of nature, which exact moral responsibilities from us. There are no 'absolute moral responsibilities' that exist independently of actual moral communities and their rules. The only possible moral orders are those emerging from human communities and human rationality.

Similarly, our 'rights' are no more than those claims we can make good within a social/moral order. Rights are therefore conditional upon the success of a claim or its enforcement through agencies lodged in the social order. Thus, there are no 'natural rights' independent of a claim and its success; not even a right for any creature, including man, to exist. It follows that what we call 'animal rights' are in fact human rights in the form of socially-prescribed liberties and restrictions which order our treatment of other creatures.

IMPLICATIONS FOR POLICY

The existence of other species, as a natural fact, does not, in itself, have any implications whatsoever for our attitudes or policies towards them and their preservation or non-preservation. To repeat, species themselves have no 'natural' rights that immediately place restrictions on the ways in which humans might deal with them. There are thus no 'inherent rights' for species to exist, for example; and whether or not a member of a species, or all the members, live or die is not, ***in itself***, a fact of any ***moral*** significance at all. Questions of moral significance in such matters only arise when human actions in relation to other species, and the motives behind the actions, have implications for, or effects upon, human interests and human moral values. So, as we have seen, human moral systems usually forbid treating animals cruelly or wantonly because of the effects on human character, but nevertheless allow other species to be used (kindly) in various ways, including killing or even extinction, where that will serve significant human interests.

It is not difficult to think of situations where the overwhelming majority of mankind would welcome the extinction of a species and strive to bring it about such as eliminating smallpox, or tuberculosis, or malaria; or the hosts and carriers of such diseases such as mosquitoes, or the snails that cause schistosomiasis, or the tsetse fly that carries trypanosomiasis, and so on. This is not to say that there may not be arguments of a utilitarian kind for preserving such species for certain purposes – scientific study or experimentation, for example – if doing so might serve some larger or more distant purpose of benefit to humanity. But 'benefit', in the broadest sense, is the test, not survival or preservation for its own sake. 'Benefits', as I have argued, can only be construed anthropocentrically to include spiritual or aesthetic benefits – such as the sheer delight for many of contemplating unspoiled landscapes and other creatures – as well as benefits of a more practical or material kind that may flow from transformation, use and, sometimes, destruction.

The fact that men and women assess 'benefits' and costs differently in relation to their individual and group values and beliefs is one of the brute facts of human life. But such disagreements or conflicts can only be dealt with through the institutions we have at hand for these purposes. In the next section, the discussion moves on to consider how the ethical systems of Western civilisation, and the varying values represented in them, have sought to come to terms with the issues of environmental protection and the continuing welfare of mankind.

PART TWO

The Preservationist Ethic & Western Civilisation

CHAPTER 6

The Emergence of the Environmentalist Movement

One conclusion of the foregoing arguments is that no proponent of an environmentalist or preservationist policy can claim a privileged ethical position simply ***because*** it is preservationist. Its claim to legitimacy can rest on no other grounds than that it serves subjective ***human*** values of a preservationist kind. And, conversely, a policy that advocates transforming landscapes, using natural resources, and perhaps risking loss of species, is not ***necessarily*** evil or irrational simply on those grounds alone. It may equally claim legitimacy on the grounds that, on balance, it serves a variety of human interests and values.

This is no more than a statement of the common sense position of most people once the metaphysical, 'objectivist' and untenable side of the environmentalist ethic has been swept aside. But policy must take account of the increasing attachment of more and more people to preservationist environmental values. How the growing importance of this commitment to the environment and to preservation is to be accommodated in Western societies and reconciled with competing values, or opposed commitments – let's say, to 'development' – is one of the outstanding ethical and institutional issues of the period we are living through.

Although the word 'environmentalism' is of recent currency, conservationist sentiments and notions of environmental stewardship and protection of species are very old. However, within the last thirty years or so, environmentalism has recruited these sentiments to a powerful, scientifically-informed and sophisticated movement arousing strong moral commitments among a great many throughout the world.

To say that 'we are all environmentalists now' is very close to the truth, but it covers a wide spectrum of commitments and different stances on particular issues. For this reason, one cannot give a single or simple description of 'the' environmental movement. It comprises a variety of positions with much overlap between them.

In the developed, rich and well-educated nations of the West, such as Australia, the mass of the people are aware of, and concerned about, environmental degradation and possible future threats. Environmental issues are quickly publicised and widely discussed. This mass adherence may not always be well-informed about the complexity and scientific uncertainty attending many environmental matters, but its desire for environmental protection and improvement is unquestionable. This desire can be quickly marshalled into a potent political force, as governments are well aware.

Beyond this mass adherence to environmentalism, the movement is carried forward by a variety of segments of varying degrees of sophistication and kinds of activism, with some committed to local or limited objectives and others deeply involved on a national basis on a range of issues. It includes those who are scientifically well-informed and careful and conservative in their recommendations, and those who are less so. It includes some with special interests to promote under an environmentalist banner and some whose commitments are unpolitical and dispassionate. There are also those whose environmental attitudes are influenced by the extent to which environmental policies may directly affect their interests in ways which may not be true for the great majority – such groups as farmers, fishermen and miners, who have a direct stake in access to land and sea and resources. Finally, there is a group of activist associations which attach an environmental critique to a broader political and economic agenda aimed at radically changing the institutional structure of Western capitalism – a group sometimes identified as the 'deep greens'; although this description might be unfair to some who feel deeply about environmental matters while not wishing to attach it to an ideological attack.

It is my contention here that our environmental problems will best be solved within a mainstream ethic that tries to accommodate and optimise both environmental amenity and the preservation of a vigorous and productive economy operating in the capitalist mode. This requires a framework of law and principles oriented towards finding the best kinds of trade-offs between these basic objectives. The last part of this study makes some suggestions about the principles that might guide such a search.

RADICAL ENVIRONMENTALISM

It is also my contention that one of the main drawbacks to reaching such an accommodation are forms of environmental activism that are absolutist and uncompromising, or which mislead and exaggerate in order to gain publicity or to serve political agendas separate from environmental protection for its own sake. Such activism appears outside the mainstream environmental movement and composes what I describe as 'radical' environmentalism. It is essentially ideological in character and there are two main elements to the ideology, either of which may be emphasised, depending upon the particular organisation concerned:

(i) the linking of an environmentalist critique with a well-established anti-capitalist, anti-market ideology, and

(ii) the identification of virtue with unconditional reverence and care for the natural world.

For some, it has become an ideology offering an alternative way of life with a comprehensive program. It is 'ecological' in this latter sense of purporting to give an all-embracing account, at least in outline, of the interconnections between natural ecology and human institutions, the whole joined together by a preservationist ethic that links concern for nature with a definition of virtue and a reformed way of life eschewing 'consumption' and 'growth'.

This has proved attractive to the idealistic young, whose enthusiasm has been recruited to what they see as progressivist organisations, such as Greenpeace, which combine aggressive and dramatic actions, adventure, defiance of authority, highly professional public relations and marketing, and opportunities for self-sacrificing service that captures their imaginations.

In making protection of the environment and preservation of species conditional upon the elimination of 'exploitative capitalism', the powerful appeal of environmentalism to broad segments of the population disarms opposition to many actions whose primary objectives are politically ideological and anti-capitalist rather than purely environmental, but which claim legitimacy under an environmentalist cloak. But such disarming would not succeed unless there was already a substantial reservoir of disinterested, ethical opposition to environmental transformations and threats to species of a non-political kind. It is this widely-disseminated ethic, characterised by disgust with pollution, the taking of aesthetic and spiritual pleasure in natural environments and the presence of a profusion of birds and other animals, feelings of awe in the presence of natural phenomena, and concern for the interests of future generations, that sits uneasily with an equally widely disseminated development ethic. The latter, which we might also describe as a 'producer' or 'growth' ethic, has dominated the West for 250 years or more, and is so familiar to us as hardly to need extended characterisation. Capitalist market economies exemplifying this ethic within politically liberal societies, have been engines of material abundance, technological progress, increasing health and longevity, as well as the loci of formidable scientific and humane civilisations.

For those who wish to preserve the best features of free market economic efficiency and abundance, political liberalism, and environmental amenity in the widest sense, the decoupling of environmentalism from the ideological attack on capitalism is an issue of great importance. Similarly, any defence of capitalism that entails devaluing environmental amenity and not taking pollution seriously will surely, and deservedly, fail.

The linking of capitalism to the supposedly inevitable consequences of environmental despoliation, to an empty, consumerist and alienated way of life, and to the exploitation and impoverishment of underdeveloped nations, are the main elements of the ideological attack on capitalism that emphasises environmental issues. The progression of such a critique of Western civilisation, when joined with a quasi-religious celebration of nature and preservationism, is nowhere better represented than in the transformation of important elements of the Christian churches and Christian attitudes towards economic growth and the natural world over recent years, to which we now turn our attention.

CHAPTER 7

Christian & Western Attitudes Towards Nature

Max Weber, in *The Protestant Ethic and the Spirit of Capitalism,* is the most famous of those who have associated Christianity with the rise of capitalism. But he is not the only one. Capitalism and Christianity have been by others as, respectively, the economic and ethical pillars of Western civilisation. Also, given the importance of Christianity in shaping the Western ethical tradition and its continuing influence, the evolution of official Christian thinking in relation to environmental matters assumes considerable significance, and this is one aspect of the discussion which follows.

It is not surprising that Christianity has been attacked by radical environmentalists on the grounds that it has celebrated man's 'domination' of the earth and its creatures and has encouraged a capitalist-exploitative attitude towards nature.

In a scholarly rebuttal of this construction of Judaeo-Christian doctrine about nature, David Elder (1994) demonstrates how it ignores many Biblical passages and Christian doctrines which stress environmental stewardship and love of nature. The attack, Elder argues, in targeting 'Christianity's huge burden of guilt' for current ecological problems, is thoroughly mistaken. Whatever the sources of any arrogant and exploitative attitudes towards nature in the West, they cannot be sheeted home to the Judaeo-Christian tradition which is notable for its balanced and sympathetic treatment of the natural world.

A broadly compatible analysis is to be found in the first two chapters of John Passmore's *Man's Responsibility for Nature* (1974), where he summarises the history, including the religious history, of Western attitudes towards nature. Passmore notes, inter alia, that the Old Testament view is that nature is not sacred and that it speaks of God having given man, the steward made in God's image, dominion over nature. Passmore observes (1974:10):

> The view that man in any sense rules over nature inevitably presumes that nature is not itself divine. And the striking peculiarity of the religion of the Hebrews, when we compare it with the middle Eastern religions which surrounded it, is its sharp distinction between God and nature.'

With various vicissitudes and changes of emphasis, essentially the same view survives in the Christianity of the New Testament and in the hands of most modern interpreters. Only God is sacred. Nature may be explored and understood. Man is God's steward both of God's instructions to mankind and also of nature; a stewardship that rejects wantonness but which allows nature to be used by man for man's purposes.

If this is the major tradition in Western thinking about man and nature, Passmore reminds us to take account of its supplementation by others (1974:39):

> There are, then, two important minority traditions in Western civilisation both of which think of men as having responsibilities towards nature. The first is, in feeling, conservationist. It emphasises the need to conserve the earth's fertility, by culling and pruning and good management. The second is rather bolder: it looks to the perfection of nature by man, but a perfection which always takes account of nature's own resources and of what man has already achieved in his civilising of the world.

These are quite consistent with the core biblical and Christian conception of the world as God's precious gift to mankind, a gift to enrich the earthly life of man. Nature may thus be a well-used instrument for man's perfection, in celebration of the God in whose image he has been made; but man's purpose is not to be the subordinate instrument of nature's perfection. So, stewardship, responsible use, working with nature, conservation, and understanding nature, together with transformation and improvement of nature to serve the needs of present and future generations; these are the main ethical themes of the Western tradition in relation to nature, variously emphasised from one generation to the next and powerfully influenced by Judaeo-Christian religion. Absent from this mainstream – at least until recently – is any widespread commitment to the idea that nature is sacred and untouchable, requiring the subordination of human interests to its interests. Nevertheless, there have been, and still are, important strands of nature-worship, ecological mysticism and romantic pantheism within that tradition. Those strands are vigorous in many aspects of the contemporary environmentalist movement and within some sections of the Christian churches.

RELIGION, IDEOLOGY AND ECO-POLITICS

The environmentalist movement, including pressures for the preservation of biodiversity, represents a number of strands of thought and action in Western modernism. The preceding section characterised the ethical principles to be found at the centre of the Western tradition. It is, I argued, an ethic that accommodates and balances the transformation and use of nature for immediate purposes with long-term care for the landscape and its creatures. The right path, on such a view, is to find the best ways of serving both practical needs as well as aesthetic and spiritual impulses of a conserving kind. Both have their place, but neither is supreme.

That such an ideal balance has not always been maintained is obvious. Ravaged and eroded landscapes, extinction of species through hunting and destruction of habitats, avoidable ugliness, pollution of soil and atmosphere, prove that the ethic has been honoured as much in the breach as in the observance. Yet the observance is substantially to be found, and increasingly so, in most of the Western world. The original wilderness of Europe has, of course, been utterly transformed; but its man-made cities, towns and landscapes have not lacked enduring beauty and fruitfulness. By and large, when balanced against the achievement of a great civilisation, the natural inheritance of Western man is not entirely dishonourable. With the growth of industrial productivity and wealth, that wealth has been increasingly devoted to restoring and improving the landscape and the habitats of species. The environmentalist movement has played an important part in increasing the environmental sensitivity of every class and occupation.

But, at its extreme edges, the movement draws support and justification for radical measures from two powerful sets of ideas and dedications; one with a relatively short history, and the other much longer. The more recent one is an ideology of oppression and exploitation that emerged in Europe in the first half of the nineteenth century. The other is a form of nature worship, as old as mankind, that has gathered new strength in the last thirty years. Much of the importance of the latter lies in its success within the Christian churches in shaping Christian doctrine towards a preoccupation with environmental issues. In view of the centrality of Christian ethics within Western civilisation in shaping moral attitudes towards nature and the use of natural resources, major changes of direction in these matters by the Christian churches will inevitably have wide ramifications. The more so if its treatment of those issues is couched in terms of categories of explanation drawn directly from the ideology of oppression and exploitation, as we shall see shortly when quoting from some recent publications. When doing so, it is worth bearing in mind Professor Kenneth Minogue's ironic characterisation of this ideology. As he puts it in the conclusion to his analysis : 'The great discovery of [19th Century] ideology has been that modern European civilisation, beneath its cleverly contrived appearances, is the most systematically oppressive despotism the world has ever known'. He goes on to remark that this ideological revulsion with Western civilisation 'coincided with the British abolition of the slave trade, and by the time Lincoln signed the Emancipation Edict it was fully established. During this same period, the groundwork was laid for universal adult suffrage, and for women to free themselves from household restrictiveness to an unprecedented degree'... and that 'the most remarkable fact about ideology is its attempt to demonstrate that what by most ordinary tests – an end of hunger and the heavier burdens of labour, respect for human rights – has been a giant leap forward by mankind, is actually a monumental retardation' (1985:221).

CHRISTIAN ETHICS AND ENVIRONMENTALISM

Apart from the inclusion of an important ideological element, what is now happening is also, in part, the resurfacing of the idea of the 'intrinsic value' of nature and its creatures within the doctrines of important sections of the churches, where it is further clothed with sacredness or untouchability. The upshot is that within these churches many leaders are beginning to embrace a kind of eco-religion, in uneasy tension with traditional teachings, that is moving the centre of Christian ethics towards the 'ecological crisis'.

We have seen that the Judaeo-Christian tradition has espoused responsible care for landscapes and wild creatures. Nevertheless, Christianity, more insistently than many of the world's religions, has been an anthropocentric or mankind-centred one in which the proper care and salvation of the soul has required Christians to place their moral duties to each other, under God, above all other considerations. But the way in which this is changing is revealed by the World Council of Churches in its contemporary preoccupations with the environment. The World Council has recently published and distributed *Redeeming the Creation – The Rio Earth Summit: Challenges for the Churches,* by Wesley Granberg-Michaelson, and in it there is a profound shift in moral focus:

> The Churches are challenged to look again at the presence of Christ the Redeemer in the midst of the Earth Summit [in Rio de Janeiro in 1992] and to discover how his presence can most clearly be expressed *for the redeeming of the earth* (1992: xiv; emphasis added).

Thus Christ, celebrated by Christians for centuries as the saviour of ***mankind***, becomes now the redeemer of the ***earth***. And the same book tells us a little later on that: 'A central calling of the churches today is to develop a spirituality of creation' (Granberg-Michaelson 1992:39). It is clear from the context that 'creation' here means the earth itself. The emphasis is not on **man's** place in creation and his stewardship of the earth and its creatures, but on the 'central calling' to develop a spiritual and theological stance towards the environment as such.

In the second last chapter of the book, 'Redeeming the Earth', this movement of the Christian churches away from a man-centred view, because of the 'ecological crisis', is plainly stated:

> A general ecumenical consensus has emerged in recent theological attention to the ecological crisis which affirms humanity's responsibility to care for and live in harmony with God's gift of creation. Older views attempting to give biblical sanction to humanity's unbridled right to 'subdue' the earth in any way necessary have been largely rejected in favour of fuller theological perspectives underscoring biblical affirmations of the ***intrinsic goodness*** of creation (Granberg-Michaelson 1992:51, emphasis added).

The careful language mutes the core message that the use of the world for the welfare of mankind should be rejected in favour of a morality centred about our duty to affirm, or protect, 'the intrinsic goodness of creation'. The fact that 'creation' includes some evil (disease organisms that plague mankind, for example) is passed over, but there can be no doubt that the moral centre of gravity has been shifted towards our duties to the earth, rather than to one another.

This shift is also associated with a range of political and economic views that support a new ethical emphasis on asceticism in the service of the earth and the relegation of human aspirations for growth and progress to the backburner.

One symptom of these new directions is the establishment of environmental lobbies within the churches and the emergence of doctrine and liturgy with an environmental focus. One such is The North American Intercontinental Coalition on Religion and Ecology. It held an Intercontinental Conference on Caring for Creation at which

Baha'i, Buddhist, Hindu and Islamic, environmentally-relevant texts were read along with Christian and Jewish ones (Bandow 1993:15).

We have already spoken of the Christian duty of stewardship for the earth and its creatures, along with the complementary duty to use the earth and its resources for the welfare of mankind. Both obligations have received varying emphases at different times, and in different places in the Bible itself. Yet never before has authoritative Christian doctrine swung so strongly towards the **subordination** of human interests in favour of the environment. There are now tendencies at work which elevate the sacredness of the earth over human welfare.

Doug Bandow, an analyst at the Washington-based Cato Institute, who has taken an interest in these world-wide Christian movements, quotes Catharina Halkes, Emeritus Professor of Feminist Theology at the Catholic University of Nijmagen, in the Netherlands, as saying: 'The image of the world as the body of God belongs more to our time and is closer to the changing reality than that of the Kingdom of God' (Bandow 1993:6).

It is a short step from this to outright paganism, and easy to show that paganism, in the form of an eco-religion, has a firm foothold in the Christian churches. The critical doctrines are already there; the earth as the body of God; the earth as sacred and therefore to be worshipped in its own right; the image of mankind as *wounding* the earth; the American Unitarian Church's Rev. Lesly Phillip's talk of 'the growing awareness of the need to honour and heal Mother Earth....' (Bandow 1993:7)

But the ramifications of the eco-brew within the Christian churches do not end there. Environmentalism-as-religion has become the linking theme, the bridge, that legitimises Church lobbying and activism on a variety of political, social and economic issues. The Rio Earth Summit was the catalyst. Thus, for the World Council of Churches, 'Rio altered the environment and development dialogue fundamentally, linking poverty, equity and social justice with the achievement of sustainable development' (Granberg-Michaelson, 1992:vii).

But it is clear that a 'sustainable' world will be one which turns its back on economic growth and which settles for what is likely to be both a uniform and Spartan lifestyle: 'The real quest has to be a style of living which can be adopted by all countries, rich and poor, and consistent with the environment and the planet (Granberg-Michaelson 1992:15). Consequently, 'The whole notion of progress, economic growth and industrialisation, with escalating affluence, is the root of ecological destruction and the continuing impoverishment of millions' (Granberg-Michaelson 1992:14).

In this, elements in some of the Churches are aligning themselves with the kind of no-growth, population-control, environmental extremism that claims that progress is incompatible with environmental protection and that industrialism means impoverishment. All of this ignores the clear facts of recent history that industrialism has significantly reduced poverty wherever it has taken hold, that growing affluence takes mankind before too long to a point where affluence itself is increasingly defined as environmental amenity, and that technological progress, discovery, and new knowledge have been responsible either for the growth of available resources or the use of resources in increasingly efficient and saving ways. It also ignores the remarkable differences, in these environmentally-relevant respects, between democratic, market-based capitalism and the environmental destructiveness of the command-and-control economies of Eastern Europe or the dictatorships of Africa.

Such blindness is matched, also, by uncritical acceptance of 'global warming' as an established fact and the result of economic growth when in fact evidence has been available for some time that such simple conclusions are highly doubtful. The 'damage to the ozone layer' is likewise quickly accepted as due to the release of chlorofluorocarbons, in the face of much conflicting evidence.

But all of this is nevertheless used as grist for a mill that steadily grinds a new grain – a new 'earth morality' based on 'A love that embraces earth, air, fire and water, yearning to bring forth a new creation' (Granberg-Michaelson 1992:68-9).

It is signalling a major change in the core ethical tenets of Christian doctrine. It points, too, to a general loss of direction that is making the Christian churches vulnerable to a desperate and unreflective attachment to new causes and the propagation of highly simplistic views in order to win back lost audiences.

OPPRESSION, LIBERATION AND THE ECOLOGICAL 'CRISIS'

As we suggested earlier, these new ethical directions in the Christian churches have strong links with ideological constructs whose natural history begins in the first half of the nineteenth century. We have come to know its manifestations in the churches in this generation as 'liberation theology'. It is important to understand that environmental and preservationist action within the framework of liberation theology does not have the simple aim of environmental conservation and species preservation for their own sakes. Those objectives are seen as consequential to what is in fact an economic and political agenda of radical social reconstruction aimed at eliminating the sources of oppression and environmental destruction; sources identified, of course, as the capitalist system and its attendant multinational companies of the world and, especially, of an exploiting 'north' – which means Western civilisation, whether in its European, North American or Southern hemisphere manifestations. When those 'exploiting' forces are destroyed, the causes of environmental despoliation will disappear as a necessary consequence.

This liberationist political orientation finds constant expression in Granberg-Michaelson's *Redeeming the Creation* (1992):

> Churches have been empowered in their ability to denounce the reigning powers of injustice and destruction (p.44).

> Perhaps the most encouraging theological development from the Earth Summit were the signs that a new integration was emerging between the classic concerns for liberation and the crucial importance of theology of creation (p.57).

> We have come inevitably to the conclusion that the prevailing system is exploiting nature and peoples on a worldwide scale and ... why it is extremely urgent that we as churches make strong and permanent spiritual, moral and material commitments to the emergence of new models of society... (p.71).

> We affirm the indivisibility of justice to the environment and social justice. The Earth Charter must clearly recognise that environmental destruction and injustice have systemic causes such as the dominant development model itself with its emphasis on capital intensive industrialisation (p. 83).

However, the drive to sacralise the earth itself, to affirm its 'intrinsic goodness' and the respect which it is owed, serves the function, whether intended or not, of adding to this politically-oriented program an evangelical and 'disinterested' moral element that reinforces and energises the movement. Although ultimately mystical and romantic, this latter aspect also helps de-legitimise the 'anthropocentric' focus of traditional Christianity with its emphasis on the dependence of mankind and civilisation on the resources of the earth and the duties of men to each other, considered not as a collectivist political program, but as questions of individual responsibility and good faith.

CHAPTER 8

The Ethics & Politics of Environmentalist Deception

THE BELIEF SYSTEM OF THE 'DEEP GREEN' ETHIC

The evangelical elements in the environmentalist movement, especially when combined, for Christians, with the kind of theological underpinning we have just examined in *Redeeming the Creation,* meet some of the conditions for the *conversions,* or 'transvaluation of values', that change the structure of preferences and incentives of large numbers of people. If and when conversion takes place, the zeal of the reforming ethic so generated becomes a potent proselytising and political force.

But, for both the religiously inclined and for those of secular bent, in order to change preferences in the required directions, beliefs have to be changed. Changing beliefs in turn requires that an interconnected set of propositions have to be accepted. I have quoted extensively above from Granberg-Michaelson's *Redeeming the Creation* because it lays bare, very simply, several of the *secular* themes and presuppositions which compose such an interconnected position.

The core is an apocalyptic vision of world crisis, an absolutist environmental ethic and an ideological critique of liberal, market economies. The key propositions underlying this vision are as follows:

(i) The natural world and species are under severe and immediate threat; there is an ecological 'crisis'.

(ii) The environment and other species are intrinsically valuable and protecting them is an absolute responsibility.

(iii) Human population growth, expanding human consumption, and spreading industrialisation are the fundamental causes of pollution, exhaustion of resources, environmental degradation and threats to biodiversity. There is therefore an immediate moral responsibility laid upon mankind to eliminate these causes.

(iv) Underlying these immediate causes of the crisis and driving the whole process is a malignant engine of destruction and exploitation identified as liberal capitalism and the market.

(v) The evidence for the crisis is apparent in that–

- (a) there is a dangerous hole in the ozone layer due to the release of chlorofluorocarbons by human action;
- (b) global temperatures are rising as a result of the release of 'greenhouse' gases because of industrial and agricultural activities, etc., with catastrophic long-term implications;
- (c) thousands of species are becoming extinct each year because of destruction of habitats for forestry, agricultural, industrial and domestic purposes;
- (d) mankind is rapidly approaching the 'limits to growth' because of the depletion of mineral and natural resources, which will soon be exhausted.

(vi) Accordingly, decisions will have to be taken now, by national governments and international convention, to reverse present trends and to ensure that growth is 'sustainable'.

(vii) Sustainable growth entails establishing more community control over economic activities.

(viii) A market economy takes no account of community-environment interactions and interdependencies; and market failures and externalities, as the major causes of despoliation, can only be eliminated or mitigated by government regulation.

The foregoing is an 'ideal type' characterisation of the blended belief structure and ethical position that sustains environmentalism as a radical movement. I am not suggesting that the 'facts' could not be added to or varied, that every environmentalist subscribes to every proposition, or that all who want to conserve or preserve are to be identified with it. But I would argue that something very like this set of beliefs would be substantially found in many present day movements and their active adherents.

Accordingly, the validity or otherwise of these core beliefs is crucial in justifying and reinforcing environmentalism as an ethical stance. For, given the untenability of the 'intrinsic value', sacralist and ethically-absolutist foundations that we have already scrutinised, the scientific status of the remaining propositions and predictions supporting environmentalist movements becomes quite crucial.

RAISING THE ALARM

What we find when we examine the various factual propositions behind the beliefs is not plain statements, but highly coloured ones offered in alarmist terms. It is, of course, quite usual for those impatient to change the status quo to characterise it in disturbing and challenging terms. Thus, the abiding *themes* accompanying the statements coming from many environmentalist movements, in addition to the constantly repeated call for the ethical subordination of human interests to biocentric interests, are *urgency, exaggeration, omission of contradictory evidence,* and *certainty*. Action cannot be delayed because the situation is so serious and widespread *now*; and the issues and courses of action to be taken are not complex, probabilistic and uncertain, but clear-cut and predictable.

Typical of this sort of approach is a leaflet on biodiversity prepared and distributed (with the aid of a grant from the Victorian Environmental Education Council and with sponsorship from the Billiecart Clothing Company) by the Australian Conservation Foundation (1992) (emphases added):

> The expanding human population with its *voracious* appetite for land and natural resources, *now* threatens *the very fabric* of life on earth.
>
> All species have *an inherent right to exist*.
>
> It is important that humans *relinquish the anthropocentric view* of nature that places our species above all others.
>
> Climate change caused by the Greenhouse Effect *will impact drastically* on species trapped in shrinking or altered habitats.
>
> Ozone depletion may destroy fragile corals ... and ... reduce plant growth which may lead to *increased global famine*.
>
> We are losing a priceless heritage often without knowing what we had to start with. [*sic*]

FALSEHOODS AND FEARS

To compare what some prominent environmentalist individuals and organisations have said and predicted with what in fact is the case and with what in fact has come about, is to be struck with their cavalier ethics. There are many examples where scientific caution and stress on the complex nature of environmental phenomena have been swept aside in favour of alarming pseudo-facts and firm predictions based on flimsy evidence.

The *locus classicus* of this technique, published in 1972, is the Club of Rome's *The Limits to Growth*. On the basis of computer modelling, it predicted that continued economic growth would rapidly exhaust natural resources, leading to economic collapse and global catastrophe. Needless to say, the implicit villains were headlong capitalism and free markets.

In the words of one reviewer at the time, the book was described as 'empty', 'misleading', and a 'rediscovery of the oldest maxim of computer science: Garbage In, Garbage Out' (Salmon 1992:54-6). Nevertheless, *Limits* was endlessly recycled as environmental gospel, with profound and lasting effects on public perceptions of environmental threats.

Similarly, in a reminiscence on the publication of Rachel Carson's enormously influential *Silent Spring* (1963), Bruce Ames and Thomas Jukes (1993:36-7) recall how: 'The book starts with a romanticised vision of a world in harmony, followed by a horror story of an "evil spell that settled on the community: mysterious maladies swept the flocks of chickens; the cattle and sheep sickened and died ... Children ... would be stricken and die within a few hours ... The few birds seen anywhere were moribund... and could not fly ... a white granular powder... had fallen like snow upon the roofs and lawns, the fields and streams".'

The powder, of course, was DDT which, as Professors Ames and Jukes go on to point out, actually saved tens of millions of lives. Condemned as a significant carcinogen, it has been shown not to be so; and there is uncertainty about its supposed effect in thinning the eggshells of birds.

Silent Spring and *Limits to Growth*, and their influence despite their falsities and inaccuracies, are typical of much of the material that helps form the belief systems of many of those who become environmental zealots, as well as the attitudes of ordinary concerned citizens wanting to protect their environments against what, on the misleading evidence they are given, seem powerful threats.

Ultimately, the significance of these two deeply flawed books rests not merely in their lack of truth, but in establishing a *mode* of politico-environmental debate whose apocalyptic sensationalism has remained a model to this day. The technique is one of making a startling and foreboding claim, without qualification and on flimsy or incomplete evidence, to draw unjustified conclusions, and then to use these conclusions as the basis for a call to political action. This, without putting too fine a point on it, is less than honest or, to be charitable, at least disingenuous. But its value, for those whose fortunes are tied to radical environmentalist causes, is in its public and political consequences; first in recruiting public alarm and then in using this alarm to pressure governments into hasty regulatory action which has two effects valued by those attached to the ideological belief-structure I have earlier described:

(i) it enhances their influence and prestige;

(ii) it moves more economic and social power over industry and commerce to the governmental centre.

That all of this may have costs and unintended consequences that damage a variety of legitimate human interests, or that reduce environmental amenity further down the track, is ignored. There are several examples, of which the DDT controversy is one. As one of the cheapest and most effective pesticides, it made a major contribution to more abundant and cheaper food, to the elimination of disease-carrying insects and therefore to better health and longevity throughout the world. Insofar as it helped to raise agricultural productivity on existing land, it reduced the demand for the opening up of virgin land and wilderness. There is no convincing evidence that these benefits have been counterbalanced by cancers avoided by reducing pesticide residues in food. The effects on bird populations are uncertain, but if it does pose a threat on this score, this has to be balanced against the costs of not using it.

GLOBAL WARMING AND THE ENHANCED GREENHOUSE EFFECT

Few subjects better illustrate the points made above about the aims and tactics of many environmentalist organisations than the stances they adopt towards global warming and the greenhouse effect.

For Greenpeace, for example, global warming becomes 'climate chaos' showing the need for 'immediate action' and the occasion for calls for federal government intervention to ensure 'integrated resource planning by utilities' and the initiation of 'energy efficiency standards and labelling for appliances and buildings' (Greenpeace, 1992). Needless to say, the costs and benefits of such action are never spelled out.

For the Australian Conservation Foundation, as we noted earlier, the Greenhouse Effect 'will impact drastically on species' and on 'coral reefs'.

The gases that compose the earth's atmosphere keep the earth's temperature at livable levels by letting sunlight through and stopping its warmth being radiated back into space. This is the 'greenhouse effect' that has been with us for millions of years and which has made possible life as we know it.

The scare about global warming refers to the so-called 'enhanced' greenhouse effect, supposedly in progress because man's industrial activities are releasing large quantities of gases, predominantly carbon dioxide, into the atmosphere and causing global temperatures to rise as they 'enhance' the greenhouse effect.

Neither Greenpeace or the Australian Conservation Foundation indicates, in their materials distributed to the general public, that global warming claims are highly conjectural as to both the extent of warming, if any, and possible causes. Nor do they mention some of the possibly benign consequences that might follow warming, if it were to occur to any significant extent – such as increased rainfall in arid areas and enhanced plant growth because of carbon dioxide-enriched air.

Although some reputable scientists dissent, it is agreed by most that the earth's temperature has increased fractionally over the last century, but the precise causes are not known. Robert Jastrow (1992) concludes that average global temperature has increased by about half a degree Celsius in the last century. Nearly all of that increase occurred before 1940, whereas about two thirds of all greenhouse gases entered the atmosphere *after* 1940. Jastrow points out that most of the predictions that have been made about greenhouse gases and global warming are based upon computer models that indicate a warming of 2.5 degrees by 2050. However, satellite measurements of global temperature over the last 13 years yield a trend line that reduces that prediction by a factor of *five,* thus suggesting an increase of only 0.5 of a degree in the next fifty years or so. In any case, satellite measurements further suggest that evidence of variations in solar activity over the last hundred years may account for *all* changes in global temperature so far observed (see figure below).

Average Global Temperature Change

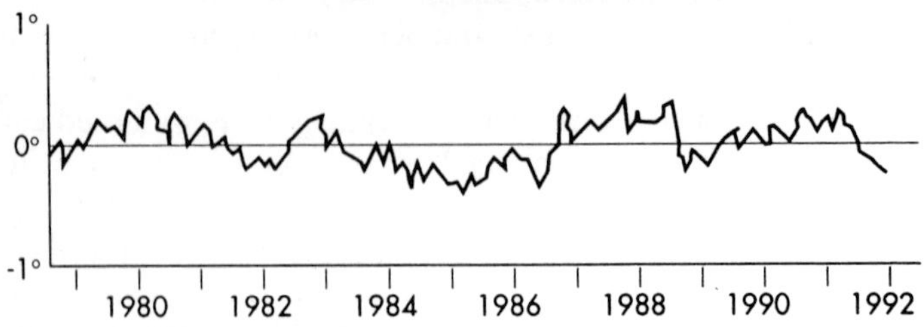

Source: The George C. Marshall Institute, Executive Alert Vol.7 No.2 March/April 1993, p6.

Presumably on the basis of the computer modelling referred to by Jastrow, The International Panel on Climate Change (IPCC) predicts a one to three degree global temperature increase by the year 2050 unless man-made emissions are cut by 60 per cent (*The Australian*, 9 February 1994). In effect, this is a recommendation for immediate and quite massive changes in the ways of life of nations, at huge cost, on grounds which are highly conjectural, on figures which vary by a factor of five, and *before* the vital facts have been established. It is hardly a model of responsibility. As American climate expert Robert Balling cautions (1993:5):

> Despite all the claims that we must act immediately, several scientists have shown that implementing policies can wait for a decade with little climatic penalty. Further, immediate action would be taken before we have the findings of hundreds of ongoing (and expensive) greenhouse-related studies around the world.

In one of the best recent summaries of the basic science of the greenhouse effect, global warming models, and the costs and benefits of limiting greenhouse gas emissions, Bate and Morris (1994:26-7) conclude, inter alia:

- There has been a measured increase in the average global surface temperature of approximately 0.5 C over the last century. A 0.5 C change is within natural variability. This measured increase is, however, likely to be attributable largely to urban warming and hence actual global warming may be very slight.
- The enhanced greenhouse effect (or 'global warming') is a highly debatable theory which does not correlate well with the empirical data.
- The models used to predict global warming, although very complex, do not include all possible, or even important, variables. Their predictions are not borne out when empirically tested.
- Temperature is more highly correlated with solar cycle length over the last century than with changes in GHG [greenhouse gases] concentration.
- The effects of increased CO_2 [carbon dioxide] are likely to be beneficial as they include increased crop yields and reduced water requirements of plants.

It is scarcely believable, in the light of such findings, that the Australian government should have an 'interim planning target' of a 20 per cent reduction of carbon dioxide emission levels below 1988 levels by 2005, at an estimated cost of $40 billion (Gill 1994). And even less so that it might consider adopting the United States' proposal for a reduction to 1990 levels by the year 2000.

Global warming, then, is another issue that illustrates the illegitimate use of the tactics of false urgency, incomplete or misleading evidence, and exaggeration, to promote extensive action with major implications for a variety of legitimate human interests. In an interview in 1989, Dr Stephen Schneider of the National Centre of Atmospheric Research in Colorado made these comments about global warming:

> We need to get some broad-based support, to capture the public's imagination. That, of course, entails getting loads of media coverage. So we have to offer up scary scenarios, make simplified dramatic statements and make little mention of doubts we might have ... Each of us has to decide what the right balance is between being effective and being honest (Coddington 1993:88).

THE OZONE 'HOLE'

Ozone occurs naturally high in the Earth's atmosphere and limits the amount of ultraviolet radiation from the Sun that reaches the Earth's surface. In 1956, Cambridge meteorologist Gordon Dobson discovered reduced levels of stratospheric ozone above Antarctica. It was later hypothesised, in 1975, that the release of man-made chlorofluorocarbons (CFCs), used for refrigeration, air conditioning, aerosols and solvents, was reacting chemically with the ozone and was the cause, or an important cause, of the 'hole in the ozone'. In fact, there is thinning of the layer, rather than a 'hole', and it appears only for a brief period in the Antarctic spring in September/October.

Scientific opinion on the causes of ozone thinning and fluctuations is marked by uncertainty and disagreement; particularly as to the relative importance, for these fluctuations, of the release of CFCs, natural cycles, including solar cycles and the seasonal antarctic 'vortex', and the discharge of chlorides from Mount Erebus, directly under the 'hole'. In short, there is respectable evidence for the view that cyclical ozone thinning over Antarctica is a regularly recurring, natural phenomenon.

New Zealand scientist, Peter Toynbee, who has been studying the ozone evidence for many years, believes antarctic ozone thinning is a natural phenomenon (Coddington 1993: 86). In addition, various studies have confirmed that there is no thinning of total ozone in tropical latitudes. Yet if chemical attack from man-released CFCs were a significant factor in ozone depletion, this should be occurring.

Nevertheless, despite evidence of this kind and the great uncertainty about the significance, if any, of the CFC hypothesis, the ozone issue was quickly seized upon by environmentalist organisations. Government responses, both nationally and internationally, were swift and wide-ranging. Australia, along with a number of other countries is a signatory to the Vienna Convention and the Montreal Protocol. At the Copenhagen meeting of the signatories to the Protocol in 1992, it was agreed that CFCs would no longer be used after 1995 and other gases substituted.

Despite the fact that there is no proof that this action will have any significant effect on ozone fluctuations, and no proof that these fluctuations are having unprecedented and dangerous consequences, societies throughout the world will lose the undoubted benefits of less costly, more efficient refrigeration, air conditioning, aerosols and solvents. The costs of abandoning them are likely to run into billions of dollars (Coddington 1993:87).

IS THERE AN EXTINCTION CRISIS?

Few issues provide as much leverage for promoting radical changes to Western ways of life as the call for preservation of species. One consequence of growing affluence in the West has been increased interest in wild creatures and their habits. Nature programs on television, whale-watching holidays, and wildlife safaris, are immensely popular. Media reports of threats to species, especially the larger, more charismatic creatures, accordingly elicit instant protests and calls to action. In short, all of this is simply one aspect of the increasing value placed by more and more people on a variety of natural and environmental amenities. It is a ready source of social and political energy that may be exploited to serve a variety of purposes. Since virtually any industrial, agricultural, mining, forestry, domestic or recreational activity can be construed as 'threatening habitats', and thus biodiversity and the continuance of species, the freedom of action to engage in those activities can be quickly put at risk if the cry of 'extinction' or 'species threat' is raised.

As we have argued earlier in relation to the credibility of elements of the environmentalist movement on various other matters, the scientific status of claims that species are being eliminated or threatened is vital in assessing the validity of the claims and in determining appropriate species-preservation policies.

In seeking to answer the question of whether or not there is an extinction crisis, we once again confront the ethically-dubious technique of raising alarms on the basis of extravagant claims supported by estimates and conjectures of doubtful scientific merit.

As we saw with *The Limits to Growth* and *Silent Spring*, a single report or book can determine the direction of subsequent discussion and, despite deficiencies of fact and argument, become a gospel justifying far-reaching policies. The common feature is a few startling claims which, when accepted, colour the whole course of what follows, including treatment of issues in the media and public perceptions.

The paradigm document in the case of the 'biodiversity crisis' is the 1980 *Global 2000 Report to the President of the United States*. It forecast that between 1980 and 2000: 'Extinctions of plant and animal species will increase dramatically. Hundreds of thousands of species – perhaps as many as 20 percent of all species on earth will be irretrievably lost as their habitats vanish, especially in tropical forests' (1980:Vol.1:3).

Eleven years later we find those figures quoted in 'Biological Diversity in Australia' distributed in 1991 by the federal Department of the Arts, Sport, The Environment and Tourism (DASET), as supporting material for the development of a National Strategy for the Conservation of Biological Diversity. A variation on the figures is presented in another DASET document, *Biolinks* (1991:1) where we are told:

> The need to protect the planet's biological diversity has never been more pressing. It has been estimated that one plant, bird and mammal species and some 50 other species are lost from the earth's tropical forests every day. This estimate of extinctions would increase significantly if species lost from other equally important ecosystems such as temperate forests, coastal wetlands, and inland waters were also included.

The figure of 53 species lost every day from tropical forests, plus 'significant' [by what measures and on what evidence?] losses elsewhere, translates to more than 19,345 species lost each year. What is the evidence for these figures?

This, and related questions, have been dealt with in some detail and with scientific rigour by Professor Julian Simon and the late Professor Aaron Wildavsky in *Assessing the Empirical Basis of the 'Biodiversity Crisis'* (1993). What follows is an extended summary of their discussion and conclusions.

The estimate of 20 percent loss of species by the year 2000 is made by Thomas Lovejoy in the *Global 2000 Report*. However, 'none of Lovejoy's references contain any scientifically impressive body of experience' (Simon and Wildavsky 1993:5-6) and the tables composed by Lovejoy are based on earlier estimates in Norman Myers' *The Sinking Ark* (1979). Myers' estimates are in turn based on a 1974 gathering of scientists who 'hazarded a guess that the overall extinction rate among all species, whether known to science or not, could now have reached 100 species per year' (Myers 1979:4-5).

On the basis of this 'guess' and extrapolating 'present [i.e. 1979] patterns of exploitations' of species-rich, tropical moist forests, Myers 'supposes' that the final quarter of this century will see 'an average extinction rate of 40,000 species per year, or rather over 100 species per day'. As Simon and Wildavsky put it: 'so an upper limit for the present that is pure guesswork has become the basis of a forecast for the future which has been published in newspapers to be read by tens or hundreds of millions of people and understood as a scientific statement' (1993:7).

Indeed, the circulation to mass audiences via the daily press of precisely this kind of material, allied with the alarmism which has proven so potent in the past, continues unabated. *The Sydney Morning Herald* of 12 February 1994, reprints in its *Good Weekend* magazine an article by Professor Edward O.Wilson of Harvard University on extinction of species which first appeared in *The New York Times Magazine*. The reprint is given billing on the dinosaur-illustrated cover of the *Good Weekend* magazine with the words: 'Hell-Bent on Extinction: Can Humanity Survive Itself?'. Inside, the heading of the article becomes: 'Death Wish: Guess Which Species is Heading for Extinction ... Ours!'

The article begins with the familiar doom-laden, highly-suspect litany of impending disasters – overpopulation, exhaustion of soil and mineral resources, ozone-thinning, and greenhouse effect from excessive carbon dioxide in the atmosphere. Man, the destroyer and polluter, is characterised as 'an environmental abnormality'; which is another way of saying (absurdly for a biologist) that man is the only non-natural thing in nature. Perhaps the only way to restore 'normality' to the environment is to get rid of man.

But the major part of the article is devoted to the extinction of species and the destruction of habitats, especially the tropical forests, by mankind. We are told that: '*Mass* extinctions are *reported* with increasing frequency in *every* part of the world' (emphases added). One might ask what the scientific basis is for these 'reports'. For example, what are the 'mass extinctions' that have presumably taken place, with 'increasing frequency', in Australia, and where are they 'reported'? Then there is the further claim that: 'Vast numbers of species are *apparently* vanishing before they can be discovered and named' (1994:16) (emphasis added). It is reasonable to conjecture that as species-rich environments disappear, as is happening with tropical rainforests, some as-yet unidentified species will be lost. But 'vast numbers' is no more than a guess, as the qualifier, 'apparently', in the quotation suggests; and there is evidence (see below, Simon and Wildavsky 1993) that species loss is by no means proportional to habitat loss.

Then comes another familiar-sounding estimation: '...it is reasonable to project a reduction by 20 percent or more of the rainforest species by 2020, rising to 50 percent or more by mid-century, if nothing is done' (1994:16). Why is it reasonable? Wilson's answer is that: 'Even a small loss in area reduces the number of species. The relation is such that when the area of the habitat is cut to a 10th of its original cover, the number of species eventually drops by half' (1994:16).

This claim brings us back to Simon's and Wildavsky's paper, because the path being followed by Wilson is the familiar one that they so devastatingly criticise. On the particular point that a 90 percent loss of habitat implies the eventual loss of 50 percent of species in it, they note that this is a common 'rule of thumb' measure used by biologists. Yet there would appear to be no empirical evidence verifying the rule. On the contrary, empirical observations in Puerto Rico, quoted by Simon and Wildavsky, show that where human activity reduced the area of primary forest by 99 percent, species extinction was nowhere near the 50 percent claimed under the rule of thumb (Simon and Wildavsky 1993: 8). Additionally, a recent criticism of the 'biogeography' theory of extinction rates has been reported in *Science* (Mann 1991). Mann notes that 'only four of 22 predictions came with sufficient explanation to permit independent examination. All of the rest provide anecdotal support or none at all. Even one prominent conservationist – who demanded anonymity, explaining that 'they'll kill me for saying this' – admitted that 'the lack of data does worry me.' He then added: 'I'm absolutely sure we're right, but a gut feeling isn't much backup when you're asking people all over the world to change their lives completely' (1991:737) .

ASSESSING THE 'CRISIS'

It should be emphasised here that we are not attempting to dismiss extinction of species as a non-problem, or to suggest that it is so minor as to be of negligible importance. Concern is justified, and it would be foolish to let a dangerous situation drift without response. But the question is whether it *is* dangerous, and what we might mean by 'dangerous'. Policies of protection of species must be measured against the real size and nature of the problem. Gross overestimation or exaggeration can easily lead to extravagant, costly and altogether inappropriate responses; responses that might well be counterproductive in the long term. As we shall see later, this has already happened in Australia and elsewhere. So scientific accuracy about the facts, to the extent that it is available or within reach, is of the first importance. Only when we have this can we rationally consider the objectives to be aimed at and the institutional methods most likely to achieve them efficiently and justly; taking account, of course, of the variety (and sometimes opposition) of the human interests involved. One can understand the interests of committed biologists in recruiting public alarm in an area of intense professional and ethical concern to them. But that very concern, and a sense of urgency, can sometimes induce overreaction and loss of judgment. Fortunately, however, the scientific ethic itself is on the side of objectivity and balance, despite the occasional backsliding and misdirected enthusiasm of some individual scientists. Ultimately, it is the continuing integrity of this ethic and the dispassionate enquiry it supports upon which we must depend. As we shall now see, that dependence is not misplaced when it comes to a scientific appraisal of the rates and risks of extinction of species.

Simons and Wildavsky (1993) report how, as a result of the criticisms and questions directed at *The Global 2000 Report* and the guesses and estimates on which much of the more alarming conclusions were based, the World Conservation Union (IUCN) commissioned a book, *Tropical Deforestation and Species Extinction* (1992), edited by Whitmore and Sayer, to investigate the extent of extinctions.

'The results of that project', Simons and Wildavsky declare (1993:8), 'must be considered amazing. All the authors continue to be concerned about the rate of extinction. Nevertheless, they agree that the rate of known extinctions has been and continues to be very low'. They then quote many of the findings, only a few of which are reproduced below. But this selection gives a reliable flavour of the rest (emphases in original):

> [The World Conservation Union] together with the World Conservation Monitoring Centre, has amassed large volumes of data from specialists around the world relating to species decline, and it would seem sensible to compare these more empirical data with the global extinction estimates. In fact, these and other data indicate that *the number of recorded extinctions for both plants and animals is very small...* ' (1993:9).

> *Known extinction rates are very low.* Reasonably good data exist only for mammals and birds, and the current rate of extinction is about one species per year (1993:9).
>
> ...[If] we assume that today's tropical forests occupy only about 80% of the area they did in the 1830s, it *must be assumed that during this contraction, very large numbers of species have been lost in some areas. Yet surprisingly there is no clear-cut evidence for this...* . Despite extensive enquiries we have been unable to obtain conclusive evidence to support the suggestion that massive extinctions have taken place in recent times as Myers and others have suggested. On the contrary, work on projects such as Flora Meso-Americana has, at least in some cases, revealed an increase in abundance of many species (1993:9).
>
> ...*the group of zoologists could not find a single known animal species which could be properly declared as extinct,* in spite of the massive reduction in area and fragmentation of their habitats in the past decades and centuries of intensive human activity (1993:10).
>
> ...how large is the loss of species likely to be? *Although the loss of species may rank among the most significant environmental problems of our time, relatively few attempts have been made to rigorously assess its likely magnitude* (1993:10).
>
> *It is impossible to estimate even approximately how many unrecorded species may have become extinct* (1993:10).
>
> While better knowledge of extinction rates can clearly improve the design of public policies, it is equally apparent that *estimates of global extinction rates are fraught with imprecision. We do not yet know how many species exist, even to within an order of magnitude* (1993:10).
>
> *There are many reasons why recorded extinctions do not match the predictions and extrapolations that are frequently published...* (1993:10).

And so on, in much the same vein. The general conclusions are inescapable that, globally:

- the prediction that 20% of all species will be gone by the year 2000 has no scientific basis;
- there is no scientific foundation for the claim by the end of this century 40,000 species will become extinct each year;
- demonstrated loss of species in recent years is very low;
- estimates of species loss that depend upon correlating loss of

species with percentage loss of habitat are unreliable.

All in all, the scientifically-reliable picture is much, much less alarming than the general public and politicians are being led to believe. But, to repeat, this is no reason for complacency. It is, however, reason to be more deliberate and measured in dealing with the situation, and reason to avoid grand, comprehensive proposals and policies of unknown impact and unintended consequences that will be difficult to unwind if and when their inappropriateness becomes evident.

EXTINCTIONS IN AUSTRALIA

On a global basis, it is clear that there are formidable scientific difficulties in accurately determining actual and likely loss of species, and that recent estimates have been grossly overstated. But this is not to say that we do not have important problems in Australia that should be addressed.

It is estimated that, since settlement in 1788, Australia has lost 20 species of mammals out of 300 known species, about 100 flowering plants out of about 24,000 species, about 10 bird species out of 850, and one reptile species out of 700. Regrettable as loss of species might be, this record over 200 years or so hardly justifies the federal Department of the Arts, Sport, Environment and Tourism (DASET) characterising ours as an 'Age of Extinction', with the implication that human settlement and land use is the major culprit. On the contrary, the record of the last thirty years, save for the depredations of introduced species (about which, more below), is one of declining rates of extinction.

While scientists are reasonably confident about many instances of species extinction in Australia, absolute certainty can rarely be claimed and there are many examples of species presumed lost turning up again. Even so famous an example of extinction as the Tasmanian 'tiger' or thylacine is still the subject of reported sightings. Graeme O'Neill (1993:48-49) gives several examples, from marsupials and plants to birds and insects, of species recorded as extinct being found again.

Insofar as loss of habitat because of human action is a factor in extinction, reduced biodiversity and reduced wildlife populations, one would expect a falling off in the extinction rate as the rate of invasion of suitable virgin lands for agriculture and grazing declines, as it has done in recent years. Even with clearing, agriculture, grazing, mining and selective timber-getting, it does not follow that species will be in peril; many native species will persist under these circumstances and some will thrive more than ever, especially species (several kinds of bird, flying foxes, insects, for example) equipped to exploit open lands, gardens, orchards, exotic plants and trees, and crops of various kinds. A spectacular example of the revival of a species through deliberate human intervention is the flourishing – even to the point of menace – of the northern salt-water crocodile.

GUILTY PARTIES

In present Australian conditions, the best short-term answer to species preservation might be the paradoxical one that we should kill in order to save. As we shall see, there is a great deal of evidence that the greatest threat to our indigenous species is not the activities of man but the depradations of introduced species, especially the fox and the cat, but not forgetting camels, rabbits, donkeys, pigs, wild horses, cane toads, goats and buffaloes. But so far as the future of indigenous species is concerned, it is the fox and feral cat which represent their greatest threats. They have been, and continue to be, responsible for more destruction than all human activity put together. It is more than probable that the elimination of these two animals would accomplish more in the protection of species than any other action whatsoever; and on those grounds, should be the prime objective of policies for the preservation of species at risk.

If one asks why this has not been an important item on the agenda of environmentalist groups, or politicians, the only plausible answer is a highly political one: it does not serve *their* organisational and personal interests, even though it might be the most important single service to the preservation of species in this country. Part of the problem is that a sustained campaign to eliminate cats and foxes would alienate supporters and potential supporters, including a large cat-loving constituency and animal liberationists who oppose the hunting and killing of foxes.

As former federal Minister for the Environment, Mr Barry Cohen has put it: 'Governments at all levels tend to be reactive to the agenda set by the conservation movement and the media. As Federal environment minister from 1983-87, I have to accept some of the responsibility for the present Government's tardiness' (1992:13).

In 1991, a national conference on the 'Impact of Cats on Native Wildlife' organised by Australian National Parks and Wildlife, concluded that cats 'were widespread across Australia and the domestic variety provides a high-density reservoir of breeding animals for the feral population; are known to kill and eat more than 100 species of Australian native birds, 50 mammals, 50 reptiles, three frogs and numerous invertebrate animals; are responsible for the carriage and transmission of infectious diseases; and are threatening the recovery programs of endangered species' (Cohen 1992:10).

Attempts to restore numbers of threatened species, such as the parma wallaby, have been thwarted by foxes. When a population of 48 of the species, bred under protected conditions, was released into the wild in New South Wales in 1988, eleven of them were fitted with radio monitoring devices. Within three months they had all disappeared and the evidence indicated that they had been taken by foxes. Cohen (1992:11) quotes Dr Jack Giles, director of Scientific Policy and Research at Taronga Zoo, Sydney, and formerly assistant director (Wildlife) of the New South Wales Parks and Wildlife Service, as saying:

> Enough research and evidence is available to start taking action against the fox. Most extinct Australian mammals were in the 50g-5kg range and lived in mainland Australia below the tropics. Foxes do not live in Tasmania and are absent from much of the tropical north and the fauna is virtually intact in these areas. I am prepared to bet that foxes are the final solution to several endangered species in Australia ... Getting rid of foxes may save a lot of wildlife quicker than any other action.

There is important experimental evidence to support this. For whatever reasons, many of the native animals that used to live in the Adelaide Hills have disappeared. In 1969 work began on a project in the Hills called 'Warrawong Sanctuary' on 14 hectares of land that used to be a dairy. The land was planted with trees and shrubs, creeks and pools constructed and the whole area surrounded by a fox-and-cat-proof fence. The animal species indigenous to the area were re-introduced.

As John Wamsley of Earth Sanctuaries reports:

> Warrawong Sanctuary has been fox and cat free for ten years. Over that time all the [native] animals introduced have thrived ... The only difference between Warrawong Sanctuary and any other bit of Australia is simply that Warrawong Sanctuary is fox and cat free (1993:2).

Wamsley goes on to conclude:

> Australia did not lose its wildlife through farming and grazing. Australia did not lose its wildlife through mining. Australia did not lose its wildlife through land clearance. Australia lost its wildlife through foxes and cats (1993:2).

From all of this we can draw four main conclusions:

(i) Claims that there is an immediate and massive global crisis of threats to biodiversity and preservation of species are scientifically unsustainable and alarmist; there is time for more scientific investigation and for moderate and cautious protective measures where they are shown to be necessary.

(ii) In Australia, the major threat to preservation of species, especially indigenous native animals and birds is, overwhelmingly, the depradations of the cat and the fox. In general, and save for special circumstances where the contrary can be demonstrated, continuing human industrial, agricultural, forestry and mining activities are, by comparison, negligible threats to species.

(iii) The most important impediment to rescuing threatened species in Australia is the reluctance of governments and many activist environmental organisations to inform the public of the importance of the cat and fox menace, for reasons which appear to have more to do with special interest politics and activist tactics than objective concern with the preservation of endangered indigenous species.

(iv) The clear and urgent direction for governmental and private action for preservation of species is to concentrate on eliminating, or at least containing, the damage being done by cats and foxes.

REPRISE

In earlier chapters we described the various elements composing the 'environmentalist ethic' and belief system. The credibility of the belief system, we argued, is crucial to the success of environmentalism as an institution-and-culture-changing movement or morality. From acceptance of the beliefs, indignation, outrage, fear, and the re-ordering of preferences follow. It is these emotions seeking discharge, and re-aligned preferences seeking realisation, which provide the movement with the energy and dynamism required for cultural change, and which find expression in the mobilisation of public opinion and the directing of pressure upon the organs of government.

The brew becomes more complex, and perhaps more potent, if the beliefs and the emotions which attend them become attached to an existing ideology – such as the longstanding theories of capitalist exploitation and oppression – which are already firmly established in some quarters; and, even more so if this complex is integrated, for some, into a metaphysical or religious world-view about intrinsic natural values and the sacredness of the Earth.

Our analysis so far leads us to these conclusions:

1. The 'intrinsic value' thesis and its derived 'biocentric' ethic lack empirical support and cannot, therefore, imply any particular moral responsibilities for mankind so far as the treatment of creatures and natural objects is concerned.
2. Insofar as it is claimed that the 'sacredness of the earth' is compatible with Christian doctrine, this may be criticised on two grounds–
   (i) that it finds little support within orthodox interpretations of Christian doctrine, and
   (ii) it depends, ultimately, upon the mistaken view that value and moral imperatives are to be found in natural objects.
3. Human values, the rules of human moral communities, and human conduct, comprise the *only* subject matter of ethics.
4. The affirmation of massive global environmental crisis on several fronts is a gross overestimation and exaggeration. Serious environmental problems are predominantly regional or local rather than global. Thus, on the best available, scientific evidence, the central propositions of the belief system on which the radical environmentalist ethic depends are unsustainable.
5. Within Australia, apart from local or regional problems, mainly to do with forms of pollution and land degradation, there is no nationwide environmental crisis of overwhelming urgency, but there is cause for concern and prompt action about the threat to wildlife posed by cats and foxes, in particular.

The pressing need is not for wide-ranging and hasty regulation, but for sustained thought about the kinds of institutional arrangements that will best protect biodiversity and environmental amenity in the long term.

This points to the importance of answering the environmentalist charge that democratic, market economies are incapable of providing the institutional means for maintaining and improving environmental amenity and that this can only be done by placing more and more power over private action and property in the hands of governments. It is to these issues that we now turn.

PART THREE

Implementing the 'Mainstream' Ethic

CHAPTER 9

Reconciling Human Interests & Environmental Protection

The dissection and criticism of the 'biocentric', 'intrinsic value' ethic and the 'eco-religious ethic' leaves the 'anthropocentric', or mankind-centred ethic, as the 'mainstream' value system and the only defensible foundation for environmental policy. It can be summarised as follows.

Most people are concerned about threats to the environment and threats to the survival of our native animals and birds. Most people want to protect them and they also want to have clean air and waters, clean beaches and a natural landscape that will retain its beauty and its capacity to enrich our lives spiritually and aesthetically. They want environmental problems, where they exist, to be scientifically and objectively assessed and the costs and benefits of remedying them made clear. But they recognise that a comfortable way of life with civilised amenities and pleasures entails drilling for oil, mining the ground, ploughing and planting, cutting trees, making things in factories and dealing with the wastes that come from production and consumption. In other words, there are unavoidable tradeoffs to be made, and the problem is how best to make them. Sometimes, it will mean disturbing and changing some aspects of the environment, and the trick is to keep this to an acceptable minimum. But people define 'acceptable' differently and people vary in the costs they are willing to bear; and this is where some problems of differing values emerge.

This mainstream' view is 'anthropocentric' or mankind-centred ethic because the focus is on *human* interests, on what is good for mankind, environmentally speaking. But this does not mean that it is a narrowly practical perspective; it takes account of man's spiritual and aesthetic needs too; and this, of course, can include valuing environmental protection.

This ethic, especially in its currently vigorous insistence on environmental protection, requires that the liberal and democratic institutions of Western societies, such as Australia, should find a solution superior to the essentially statist proposals of the deep greens. This is a formidable challenge. It means thinking 'ecologically' about the interfaces between human action, especially economic action, and nature.

Somewhat surprisingly, the problem with much that issues from the 'deep ecology' movement is not that it is too much concerned with man-environment relationships, but that it is insufficiently so. What we find too often is a very simplistic view of human systems and institutions that is anything but 'ecological' in its thinking and which has little conception of the subtleties of interaction between human and natural orders.

This is all the more surprising in view of the recognition that human systems have enormous potential both to harm and benefit the viability and stability of natural systems. Given this potential, one would think that comprehensive ecological thinking would pay particular attention, therefore, to the ways in which different types of human systems interact with, and differentially affect, natural systems.

So we need institutional arrangements that are flexible and efficient in accommodating the huge and constantly changing variety of preferences and expectations that people have about economic satisfactions and environmental amenity. And this must be achieved within a framework that preserves liberty.

It is the biological sciences which tell us most about the interactions, interdependencies and feedback loops that are so important in maintaining the viability and dynamism of ecologies. It is economics which best reveals similar mechanisms at work in human action in pursuit of economic preferences and satisfactions. Both sciences are vital in explaining what happens at the nexus where human and non-human systems interact. But it is here that the deep ecologists falter. Their appeal to the political system to take over and 'manage' the interface is really an abandonment of the will to understand how, in fact, human political and economic systems do, or might, interact with natural systems, and especially how environmental and non-environmental values can be mutually adjusted.

So far as economic systems are concerned, the models available are the socialist and the free market, and their hybrids. The crucial feature distinguishing them is the extent to which power is centralised or decentralised. Socialist societies are characterised by highly centralised command and control of the economic interfaces between natural and human systems, with relevant information for decision-making being channelled through a bureaucratic hierarchy to the decision-making centre. In the free market, the distinctive features are clear and disseminated property rights, exchange under the rule of law, and the emergence of a price mechanism which delivers information to dispersed controllers of resources (holders of property rights) about changing values and thus the revelation of opportunities for personal advantage and more efficient allocation of resources, including allocation to achieve environmental and non-environmental satisfactions. Economic power and the dissemination of information are highly decentralised.

COMMAND-AND-CONTROL

The deep greens or deep ecologists support government intervention at the national level and encourage the proliferation of international environmental treaties. Leaving aside ideological agendas that might be at work here, this reveals an optimistic view of the incentives for governments and bureaucrats to promote environmental amenity, and their power to do so.

The main justification for government intervention offered by the deep greens is the need for governments to control the 'negative externalities', the environmentally undesirable consequences of private and public economic activities, such as pollution, the generation of noxious wastes, and so on. This is a subject discussed in more detail later. But their occurrence is of great importance for raising public concern and using that concern to achieve the intervention and control over private activities that fits their programs. As we have seen, this is why supposed global threats assume such importance for radical environmentalists. The larger the negative externalities, the more potent the accusation of 'market failure'.

And we must remember that some elements of the green lobby have a vested interest in crisis-cookery. So they are constantly starting environmental brush fires in order to keep their names, and the environmental ferment they depend upon, in the media. This gives them power over governments and therefore power over us. They also happen to be an elite who claim to know what's good for us; and their aim is to use governments and their bureaucracies as instruments for engineering the kind of world-view and ideology that I have described. And they have had stunning success, as we all know.

The upshot in this country has been a rapid drift towards a command-and-control approach in environmental matters which I believe is insidious and dangerous; dangerous not only to wider liberties and our federal system, but especially for environmental protection. Let's look quickly at some recent history.

The federal election of 1990 was held in an atmosphere of intense environmental agitation, with the circulation of the kind of misleading information I've talked about, and political parties striving to lock in 'green' votes. The domestic conditions had been created by then in which both federal and state governments had committed themselves to strong action. Externally, the Brundtland Commission report *Our Common Future* of 1987, and the mid-1992 UN Conference on the Environment, held in Rio de Janeiro, added global pressures to the heated domestic scene. The federal government entered the Rio Conference with a readiness for extensive national commitments which it subsequently made. The outcome has been a series of decisions whereby:

- agreements between the states and the Commonwealth, reinforced by the Commonwealth's eager use of its external affairs powers, have greatly increased the Commonwealth's control over the states and territories in environmental matters;
- the National Strategy for Ecologically Sustainable Development of December 1992, locks the Commonwealth, states and territories into some 384 commitments to make environmental and species preservation interventions in primary and resource industries, other forms of production, and governmental services such as public transport;
- almost at a single stroke, the governmental and administrative processes involved in access to and use of land and sea for agriculture, mining, fisheries, forestry and industry generally, threaten to become immensely more complex, legally and operationally uncertain, and costly.

In the words of Perth environmental consultant Dr Brian O'Brien, recent inter-government agreements have 'effectively nationalised the Australian environment' (1993:1).

This process continues to gain momentum in ways only too familiar to those relatively small groups who are deeply affected – miners, farmers, manufacturers, foresters but largely unknown to the general public, which remains unaware of the long-term implications, not only for their economic welfare but also, as I shall later argue, for the protection of the environment.

The general thrust of the regulatory and administrative structures erected over the last few years is towards concentrating power in Canberra, while retaining some democratic safeguards on this power. However, the mode of operation of the latter may have the effect of making the processes attending development applications much more uncertain, costly and time-consuming. For example, under some current proposals power may come finally to rest in a ministerial veto over development projects. But ministerial approval, if finally won, may then be challenged by anybody. For instance, a recent discussion paper issued by the federal Department of the Arts, Sport, Environment and Tourism (DASET), proposes that ministerial decisions with environmental consequences could be challenged by any citizen in the Administrative Appeals Tribunal. In other words, plans and proposals for major projects years in the making may finally find their way through the bureaucratic labyrinth to ministerial approval, only to find that approval challenged by any individual or group. If the challenge is upheld, the project may have to be abandoned or painfully reconstructed and re-submitted to the process again.

As recent agreements between the states and the Commonwealth proceed to the drafting of legislation to establish Australia-wide environmental protection standards, the indications are that the National Environment Protection Authority (NEPA) implementing those agreements will be empowered to formulate *national* measures of great scope, dealing with such things as water quality, noise, wastes, recycling and vehicle emissions (Hooper 1994:30).

Yet another proposal is to set up a national reserve system which could greatly expand, even double, the areas presently given over to national parks and reserves.

INTERNATIONAL TREATIES

We earlier noted the power to raise alarm and action with the cry that species are threatened with extinction. Accordingly, negotiations to formulate an international biodiversity treaty have profound implications for Australian industry, agriculture and forestry. Against the background of the High Court's interpretation of the scope of the Commonwealth's external affairs powers under the Constitution, the signing of international treaties now gives the Commonwealth hugely extended domestic powers that by-pass the states. Since domestic interests in Australia usually have little opportunity to scrutinise these treaties before they are signed, those interests might suddenly be confronted with a virtual fait accompli on matters of great substance deeply affecting them. So much so that, as Hooper notes: 'The pent-up frustration broke out last month when eight industry groups, including the Law Council of Australia, demanded full consultation by government with industry and the community ahead of treaty-making, as well as tougher Cabinet monitoring' (1994:32).

WHEN THE STATE TAKES COMMAND

Australian environmental policy is already well-advanced towards the concentration of huge powers over the environment and industry in the hands of a few ministers and bureaucrats in Canberra. The present issue is how this power will be used and its use monitored. The problem of environmental protection for democracies where property rights are more or less secure and ownership of industry mostly in private hands is different from the situation in socialist countries. We now know what has happened in the latter, and why. Even so enthusiastic an advocate of governmental and inter-governmental solutions to environmental problems as American Vice-President Al Gore, acknowledges that (1992:247-248):

> The most serious examples of environmental degradation in the world today are tragedies that were created or actively encouraged by governments – usually in the pursuit of some notion that a dramatic reordering of the material world would enhance the greater good. And it is no accident that the very worst environmental tragedies were created by communist governments, in which the power of the state completely overwhelms the capabilities of the individual steward. Chernobyl, the Aral Sea, the Yangtze River, the 'black town' of Copsa Mica in Romania – these and many other disasters testify to the severe environmental threats posed by statist governments.

To which we might add the fact that in East Germany pollution has destroyed 9 per cent of agricultural land. The Polish Academy of Sciences estimates that a third of the country's population lives in areas of ecological disaster. Hungarian authorities believe that one death in every 17 is caused by air pollution.

That these disasters are not due solely to the unique excesses of communist totalitarianism, but may still occur under more democratic regimes that nevertheless centralise industrial and environmental power in the hands of a few at the governmental centre, is indicated by the fact that Russia continues ***daily*** to spill four times the oil released in the Exxon Valdez disaster in Alaska. The countries with the greatest control over their economies and their environment are the world's worst environmental vandals and polluters.

I am not suggesting that we are approaching such a situation in this country, because ownership and substantial control of industry still remain in private hands, and it is highly unlikely that we will ever approximate the socialist situation where nobody can be held responsible and answerable for environmental degradation. Rather, the danger here is that those in government who regulate the environment are not answerable for the *costs* they impose in doing so; costs which are ultimately diffused among millions of consumers in the form of a slow progression of higher prices and lower incomes than might otherwise have been the case. In other words, there is no acceptance by the regulators of a responsibility to balance costs and benefits and no established democratic procedures for calculating and monitoring that equation. On the other hand, regulation – often hastily and unreflectively imposed – has immediate benefits for governments seen to be 'doing something' or placating special interests.

Some 'green' agitation for environmental measures frequently entails substantial government intervention in areas where governments so frequently fail. Thus, the more proactive see governments as initiators of plans to speed up environmental protection and rehabilitation through government sponsorship of 'appropriate technology' and environmentally friendly products; by government purchasing plans to favour certain technologies and products over others; by modifications to public transport; by funding of certain kinds of research and development into technologies; by training schemes; and more. All of these figure in pamphlets and literature issued by environmental organisations and are being put on the agendas under consideration by government (Greenpeace 1992).

Ben Lieberman (1994:6) gives an American example from the Environmental Protection Agency (EPA). He describes how a small research and development firm, Sunpower Inc., with a $US250,000 grant from the EPA, designed what the EPA subsequently called a 'revolutionary' new compressor for domestic refrigerators. Its claimed increased efficiency would lead to reduced greenhouse gas emissions and was reported to work well with the new ozone-friendly refrigerants currently being introduced. According to the EPA it could eventually be used in small air-conditioning systems, saving $US4 billion per year. An EPA administrator claimed that the device 'will do for the refrigerator and air conditioning industry what the microchip did for the computer'. But, as Lieberman goes on to point out, the device uses a basic idea which goes back to the 1930s. It never caught on because compressors using them were considerably less reliable than conventional compressors. Despite improvements in the new device, spring fatigue and piston misalignment are likely to ensure that it will break down more frequently, leading to more expense and inconvenience for consumers. As Lieberman puts it, 'there is nothing for EPA to crow about – other than discovering yet another way to squander 250,000 taxpayer dollars.'

Such episodes are straws in the wind pointing to the perennial problem of governments 'picking winners' and subsequently absolving themselves, when they fail, of the costs borne by consumers and taxpayers. In the process, however, the availability of taxpayers' funds and government favours that follow the intrusion of government into fresh fields, inevitably attract producers and special interests who soon become dependent upon subsidies and favours. And so, more special interests become entrenched to contribute to the creeping paralysis that such relationships inflict on the economy. When cost-benefit analysis is discarded, when the test of the market place is never applied, in the name of 'environmental protection' the national economy becomes the plaything of fads, governmental experts, and the urgings of lobbies.

A current example of faddish rushing-in without adequate prior analysis and cost-benefit calculation is recycling. This is very popular with state and local governments and already very substantial investments and running costs have been incurred. In some cases it has been made mandatory for consumers. Yet careful studies of some (but not all) examples have shown large extra costs for consumers and the generation of large amounts of hazardous wastes from the recycling process (Cato Institute 1994).

Lead can be dangerous when absorbed by the human body, and it is obvious that measures to eliminate absorption or to keep it to safe levels are sensible. But there is a point at which slight or infinitesimal risk must be gauged against large costs – costs which may hinder our capacity to deal with more urgent health matters. What has been happening with lead reduction measures indicates the power of environmental scares, when up and running, to overturn the need for such sober calculation.

In commenting upon the lead-abatement strategies of governments in the *Medical Journal of Australia,* researchers from Monash University's Department of Social and Preventive Medicine point out that it is not known what contributions leaded petrol makes to children's lead intake. Despite this, it will cost $7 billion to remove lead completely from petrol by 1995. It has also been estimated that to remove leaded paint from older houses would cost $2,500 to $15,000 for each one, with what may be only slight benefits (*Sydney Morning Herald,* 9 February 1994).

What these examples show, and what the history of environmental devastation in socialist and dictator-led countries demonstrates, are the dangers of centralised planning and control in environmental and economic matters, and the danger of such action in democracies being driven by special interest lobbies.

When such a warning is raised in democratic countries, it is frequently said that a democratic political system is an automatic safety mechanism against centralised excesses in decision-making. But this is far from the truth. A vote at a general election once every three years cannot be a vote on each and every economic or environmental decision by government. Nor do voters necessarily make the connection between environmental policies and their often distant consequences; such as economic decline or increased taxation. It might be said that it is not elections, but scrutiny and debate in the Parliament that constrains the abuse of centralised power. But, here again, the formalities bear little relation to the realities. Under rigid party discipline, especially if one party controls both the House of Representatives and the Senate, the Cabinet and the Prime Minister exercise largely unchecked power within the constraints of the Constitution – a Constitution itself significantly emasculated in its federal, and hence decentralised dimensions, by recent decisions of the High Court; and by the federal government's use of its external affairs powers increasingly to control domestic affairs at the expense of state powers. The upshot is that democracy in Australia is not at present the check it is believed to be on command-and-control environmentalism.

If we are successfully and efficiently to implement the 'mainstream' ethic in environmental matters, it is essential that the proper role of government be defined; and defined in relation to private rights and private enterprise.

CHAPTER 10

Property Rights, Markets & Prices

We have reached the stage in this country where uncertainty over the ways in which governments will use their environmental regulatory powers is inhibiting economic activity and new projects. The extent to which private enterprise can maintain and extend economic activity is contingent upon the extent to which it can predictably command the resources to do so. If governmental environmental policies are to work cooperatively with private economic activity within the overarching objective of fulfilling the 'mainstream' ethic of continued growth with adequate environmental protection, those policies need to achieve two objectives:

(i) they must not destroy private initiatives and entrepreneurship, and

(ii) they must provide positive incentives for those controlling private resources, including land, to protect environmental amenity even as they increase productivity.

To do otherwise, to depend wholly upon regulatory sanction and negative incentives to shape environmentally sensitive private actions, would be both oppressive and destructive of liberties as well as environmentally counter-productive; for several reasons – many of which have already been given but which will be added to in what follows.

SUSTAINABLE DEVELOPMENT

The slogan, 'sustainable development' is a useful starting point. It is now widely used and it sums up the compromise, or trade-off, between continuing to seek abundance and the good life while preserving environmental amenity. It succinctly states what I have called the mainstream 'anthropocentric' ethic. The problem is to make it workable.

One of the difficulties with sustainable development is that what is sustainable or what ought to be sustained are themselves by no means clear. Five thousand years ago it could have meant conserving supplies of flint. But if the flint had been conserved by using it sparingly, progress beyond the necessity to use flint might have been long delayed. Such conserving policies would not have served succeeding generations well. The same might today be true of oil and copper ore, for example. What constitutes a resource is constantly being redefined by technological progress which itself depends upon using present resources; and, if those resources continue to be valuable as the quantity available declines, their prices will rise as they become scarcer and thus less of them will be used. And so the price mechanism of markets points the way to making sense of the idea of sustainable development. As resources are used and their prices rise because of scarcity, the rewards for finding more of them, or substitutes for them, will increase. So, either new resources will be sought; more plentiful substitutes will more likely be found; or scientific and technological progress will more likely enable us to meet our needs with less – just as chip circuitry and electronic advances have allowed us to do more with less energy expenditure. Similarly, fibre optics made from sand are now replacing the large quantities of copper formerly used in cables and wires. Or, to take another example, tractors, genetic engineering, insecticides, fertilisers and other agricultural innovations now enable us to multiply crop productivity many times over while using *less* land. This not only improves human well-being, it means that land may be preserved as habitats for native creatures. Indeed, the significance of technological innovation for environmental conservation and preservation of species can hardly be exaggerated, as the following illustrations show.

A major threat to species may arise from destruction of habitat through land-clearing for cultivation. Roughly speaking, the amount of land needed to be cleared for cultivation is a function of population pressures and the productivity of agricultural techniques. At a given level of productivity per uniform land unit, as the population increases, proportionately more land will need to be brought under cultivation.

Technological progress applied to increasing productivity per land unit reduces the need for more land as population pressures mount. Not only does this mean less pressure on unused land, it may also release marginal land, presently under cultivation or stock use, for reversion to natural habitat or less intensive use. Although other factors have also been at work, this has been happening to dairying land in New South Wales. In the American state of New York, there were 24 million acres of farmland in 1880; today there are about 8 million acres. 'As a result, wildlife is more abundant and variegated today than it was at the turn of the century, and some species are making a comeback' (Goklany & Sprague 1992:14).

The United States also illustrates this trend on a national basis. In 1910, for example, the USA needed 80 million hectares of crops to feed and clothe a population of 92 million people; plus another 88 million hectares to feed horses and mules – a total of 168 million hectares. But by 1988 it needed only 70 million hectares to feed just on 250 million people, and the need for crops for horses and mules had disappeared because of technological innovation. If, in 1910, the US had followed the advice of today's radical environmentalists and frozen advances in insecticides, fertilisers, tractors, and so on, another 220 million hectares of land would have had to have been brought into cultivation to feed today's population, plus tens of millions of acres to feed horses and mules (Goklany & Sprague 1992).

The general point that emerges from this is that invention, innovation and creativity are, by definition, unpredictable and have unpredictable consequences. Who would have thought, 150 years ago, that 'rock oil', deep underground, would be transformed from a nuisance to farmers when it sometimes seeped to the surface, to become a vital agricultural resource when used as the basis of tractor fuel? Or, more recently, that the same substance would come to be the raw material for much of our clothing and other processed goods?

The human mind is the 'ultimate resource', but its energetic employment to solve human problems or to make discoveries depends to a large degree on the incentives available to encourage it. A free market system generates those incentives. The *Limits to Growth* predicted in the 1970s that supplies of silver, lead, zinc and mercury would be exhausted by the year 2000. In fact, reserves of these metals have increased; demand has changed; substitution has occurred; and economically-profitable re-cycling has been stimulated (Stavins 1993). So, we cannot predict today what sort of detailed policy now will ensure sustainable development tomorrow.

It is the dynamism, the search for entrepreneurial opportunities, and the rapid and frequently innovative response, guaranteed by markets, that constantly undercuts policies erected on a view that the future will simply be an extrapolation of present trends. Yet precisely this static, non-dynamic view is implied by a great deal of the regulation concerned with 'sustaining' environments. It leaves very little place for entrepreneurial discovery.

Implicit in the notion of 'sustainable development' is a conception of improving quality of life in all its dimensions. But, except in a dictatorship, there cannot be a ***single*** view of what constitutes 'quality of life', much less a ***stable*** view, except perhaps in terms of meeting a few basic needs. Individual preferences and the norms and values of groups are exposed to constant change and challenge. 'Development' implies this to some extent; and, indeed, the growing importance of environmental values in our own lives is evidence of this – and a function, in large part, of the easy satisfaction of the more basic needs that actual 'development' and growing wealth have accomplished.

That such things have occurred and continue should be sufficient to provoke the deeper 'ecological thinking' whose relative absence in the green movement I earlier deplored. It may be too glib to conclude without further discussion that ecological sustainability ***depends*** upon continued development through preserving the basic institutional arrangements that have brought us thus far; but it is close to the truth. The fact is that the most environmentally sensitive nations are, by and large, the most developed nations, as well as those which have retained the fundamentals of a market economy and high degrees of personal liberty.

So, there cannot be a rational policy for sustainable development except of the most general kind, limited to a few guiding principles grounded on our experience to date and putting a premium on arrangements that will permit maximum flexibility within a legal framework that is straightforward, predictable and steady. Flexibility is of the essence in dealing with a dynamic and volatile global economy where unanticipated changes are daily fare. But flexibility and adaptability themselves require the rapid generation and appropriate diffusion of relevant information to those in command of the allocation of resources, from the individual citizen, through small businesses, to the largest corporations.

As F. A. Hayek (1978) pointed out several years ago, the market, with its price mechanism, is an information-generating-and-disseminating 'discovery procedure' that indicates where and when resources might most efficiently be allocated to yield maximum satisfaction of individual and group preferences, including environmental preferences. The market rewards action to preserve, for example, when doing so serves consumer interests or values, and it is rapidly adaptable to an infinite variety of changing tastes and values. If a particular good becomes scarcer – let's say the opportunity to observe native animals in natural habitats – as the demand rises, then the market signals an entrepreneurial opportunity to develop and stock a native animal reserve that will respond to that demand and reward the supplier for doing so.

Property rights and responsible stewardship go together. Conversely, when nobody enjoys property rights in an amenity or good available for use and exploitation at no cost, then those amenities or goods may be used, or used up, ruthlessly or irresponsibly. In 1968, biologist Garett Hardin coined the term 'tragedy of the commons' (1968:1243-48) to summarise what happened to pastures communally held by medieval villages. The owners of herds had the right to pasture their animals on this communal land. Clearly, it was in the interests of each herd owner to maximise his benefits. If he restricted his pasturing to let the grass recover, there was a risk that some other herd owner would capitalise on the fruits of his restraint. The upshot was that there was no incentive to conserve or take a long-term view, with obvious and tragic consequences as the common land was steadily devastated, to the disadvantage of all concerned. It is sometimes the case, however, that local communities may evolve rules for the management and non-exploitative use of common areas that may keep them in good order and productive for hundreds of years. Indeed, the evolution of property rights is itself a case in point. The more general point is the importance of finding means for equitably apportioning benefits and for devising incentives and disincentives, which may be material or non-material and moral or social, for discouraging self-interested exploitation or destruction. It may well be that the system of property rights as we understand it is the most efficient means of achieving these ends.

In the absence of effective rules of these kinds, *all* common or collectively held property is exposed to risk. Public train carriages, streets, parks, wilderness, beaches, rivers, unowned animals or birds, the air we breathe, may be damaged, appropriated, misused or polluted when there are no individual or group rewards for conserving or protecting, or when the benefits of doing so may be captured by others.

For ourselves, we can thus begin to see that one of the greatest threats to the environment and other species occurs when ownership rights are lacking, difficult to assign, or severely attenuated by government controls. When lakes or rivers, forests and fishing grounds, wildlife and their habitats, and the air we breathe are unowned but open to access or use as common property, then they will be at risk of misuse. Certainly, strongly entrenched customs or habits of gentle and considerate use may mitigate selfish or callous behaviour, but it is an uncertain safeguard in comparison with firm property rights.

Amongst other things, property rights define the areas where owners have privileged use. Beyond those areas, the law prevents harm to others or demands compensation for harm or trespass on the rights of others. Without property rights firmly protected by law, the struggle to advance personal or group interests is transferred to the political process where success consists in recruiting the coercive powers of government to advance one's interests. It is precisely this struggle which is intensifying in Australia as power over the use of property in ways which might have environmental implications increasingly accrues to Canberra. The 'currency' which bids up the price of preferment by government is votes, or the perception by political parties that votes are under the command of special interest groups.

There are several good reasons, then, why markets more efficiently, peacefully and non-coercively serve competing interests and values through the price mechanism, and the rewards involved in meeting environmental demands. But there is more to it than simply appealing to markets to solve environmental problems in a mechanical way at least cost. Markets give innumerable individuals and groups freedom of choice and action; freedom to transact and negotiate with others in an ongoing process which is independent of the political arena, and where the only third party is not a government minister or bureaucrat exercising discretionary powers, but impersonal and impartial law.

NEGATIVE ENVIRONMENTAL CONSEQUENCES OF ECONOMIC ACTIVITY

We suggested above that the negative environmental consequences of some actions by individuals or organisations for 'common property', such as pollution of rivers, beaches, oceans, air, etc., arise because no one has a property right interest in protecting these collective or public goods. Such a situation is often characterised as 'market failure'. Economic activities which produce noxious wastes as by-products which are delivered to others in the form of polluted waters or air may not entail any costs for those producing them. Those adversely affected may not be able to identify the offender in a legally certain way and may thus be unable to claim compensation for the harm done. Not surprisingly, governments are called upon to correct such things by regulations punishing those who pollute, if they can be identified, or by preventing the installation and use of equipment that might produce such results. Such a call for government regulation arises because of the difficulty or impossibility of establishing private ownership of the air we breathe or the oceans, for example.

But is it, in fact, beyond human ingenuity to find solutions to at least some of these problems by assigning or selling property rights and establishing markets in those rights? If we could do so we would avoid the dangers of excessive and costly regulation, restriction of freedom and emergence of special interests entering the political marketplace. Russell Lewis has suggested such an approach to motor vehicle emission pollution of the atmosphere:

> One answer is to privatise roads, treat them [i.e. the roads] as the source of the pollution and make the owners liable for damages. The road owner would charge lower tolls for cars with better pollution equipment while those with no equipment would be banned. This should be very effective because research has shown that 10 per cent of cars cause 50 per cent of pollution. Charges would also be lower at low traffic times and higher at the peaks, which could be established with automatic sensors (1992:4-5).

Whether or not such a proposal would prove to be practicable, it is an example of a *mode* of thinking that seeks to avoid the dangers and rigidities of command-and-control reflexism and which reinforces the important insight that excessive consumption of natural resources of all kinds (even air) will occur if their prices do not, in some form or other, include their external costs.

A corollary to this way of thinking is 'getting the incentives right' in order to encourage voluntary and private conservation and preservation. This is particularly important for the preservation of native species of birds and animals in Australia. Policies adopted so far virtually ignore this. They depend upon command and control and the extension of reserves and parks – with two consequences.

The first is that the areas are so large as to make effective monitoring of ecosystem changes very costly. And, as we have already noted, unless something is done to stop the depradations of introduced species such as foxes, cats, rabbits, etc., there will be no relationship between enlarging reserves and more effectively protecting biodiversity.

The second consequence is that extending reserves that lock out mining, agriculture and forestry, or which resumes land already used for farming, imposes large costs without any corresponding analysis of benefits. As a report by the New South Wales Farmers' Association puts it (1993:3):

> ...if we force landowners to wear the costs of preserving endangered species but give them no stake in it, we are making those species liabilities, not assets.

The same report goes on:

> A means of bringing value to an endangered species is to give the landowners the right to harvest native fauna and flora and to then market them, including to overseas markets (1993:4).

As the report suggests without going on to draw the obvious conclusion, when species on private land become liabilities rather than assets, an incentive is created for their destruction; to follow a policy, as American farmers in a similar position put it: 'To shoot, shovel and shut up.'

Government action or regulation that changes the conditions under which land may be used, such as may occur if restrictions are placed upon use of the land to preserve species, amounts to a form of expropriation if the landowner is not compensated for the effects on income and future value of the land.

CHAPTER 11

Self-Organising Systems & Environmental Adaptation

We might begin this chapter with an analogy. One of the forces driving ecosystems to greater variety and adaptability is the dynamic interactivity of open systems. The richer and more various the organisms and resources of a given ecosystem – a rainforest, for example – the more niches (opportunities?) there are to be exploited; and, as those niches are filled, the organisms so established become ecofactors in their own right, contributing to the ever more complex feedback loops and interdependencies that characterise a self-organising system (diZerega 1992). This last notion, 'self-organising', is important. There is no master controller at work here; no eco-manager devising regulations, receiving information and dispensing decisions. Information and decision-making are radically decentralised. An evolutionary logic and discovery process is at work – organisms responding to 'information' coming from their environments and accommodating contingency by selective mutation or variation which allows species and organisms to keep pace with physical transformations in their environments, or go under. Such success as there might be for a species (and it is never guaranteed for all time for any species, including man) is 'remembered' in its genes; a resource reflecting aeons of struggle and adaptation.

But the human equivalent of biodiversity is not simply the genetic adaptations or racial variations of mankind; it is as much kinds of culture. So far as homo sapiens is concerned, culture, ways of life, moralities and value systems are the more important 'gene pools'. It is through these institutional artefacts that the history of the species confronts present environmental contingencies with the most important of the resources it can command. The significant variations of humanity are not racial but cultural. For humanity, successful adaptation is measured not merely by survival, but communally, in the persistence of **ways** of life and civilisations. But just as no species is guaranteed survival, neither are ways of life.

To pursue the analogy, in organic evolution the key mechanisms are *variation* (by genetic mutation), *selection* (of mutants in the struggle for survival) and *retention* of successes (in the genes). The human cultural equivalents are variation via *risk-taking, innovation and enterprise,* selection through *competition,* and retention through *institutionalised traditions* and their repositories. What institutions and traditions carry on is remembrance of principles of action, of ways of doing things and dealing with kinds of problems that have worked in the past. Custom and tradition in matters of science and various forms of practice are the indispensable materials of thought and reason and thus the necessary conditions of adaptation and change to meet novel contingencies. It is thus not surprising that *'Mnemosyne'* ('Remembering') was the first of the ancient Greek muses.

In a mature civilisation, various lines of development, or traditions marking the accumulated explicit and tacit knowledge distinctive of particular fields of activity, tend to come together, more or less coherently, and cooperate. For example, the ways in which scientific progress in one discipline, or amongst one group of scientists, may cross-fertilise another; or the ways in which the rule of law and its integration with property rights and markets represent principles of economic action and control of misconduct that work together to ensure smooth, predictable and less costly transactions between people. These institutions are notable, as we have earlier remarked, for their rapid information feedback loops, and their adaptability and flexibility. They also contain built-in incentives and rewards for innovation, risk-taking and enterprise; and all of this reinforced by the competitive climate which is part of the definition of open societies and free markets and which root out failures. The circle is completed by those institutions which instruct each new generation in the essentials of the culture which has made their appearance and flourishing possible; such things as families, schools, universities, libraries, customs, professional associations, and organisations persisting over many generations. In short, a 'gene pool' of cultural continuity and regeneration. The continuing vigour of that gene pool depends upon its retaining its self-organising characteristics, which in turn depends upon governments protecting their functional capacities and not seeking to appropriate them to itself by assuming their powers, prerogatives and resources. This means that government must be confined to the fine-tuning of the general rules that ensure no more than their effective and efficient interaction; of maintaining a benign 'ecosystem'.

And so, to end an analogy in danger of over-stretch, the focus should therefore be on *the processes* that characterise self-organising systems, on ways of carrying on, rather than on trying quickly and by command to reach particular ends. It is the preoccupation with achieving particular environmental ends quickly, by fiat rather than adaptive process, which characterises political thinking and command-and-control makeshift. It leads to the steady destruction of the private capacity for adaptive process lodged in private property rights, the common law, and other established institutions which have succeeded in solving problems in the past.

THE CONDITIONS OF CONSERVATION

The great danger facing good environmental policy-making in Australia is not lack of governmental action, or even tardy action; more frequently the danger is hasty and inappropriate response, particularly in reacting politically to a range of environmentalist scares in order to win votes. The focus becomes short-term vote winning by making regulations or taking discretionary cabinet or ministerial action, when the real need is steadily and predictably to build private institutional capacity, and private incentives, that will serve the environment well for an indefinite future. I will shortly make some suggestions as to how this might be done.

We have seen that the claims of imminent global environmental catastrophe and massive extinctions of species are wildly exaggerated. Nevertheless, Australia, in common with most other countries, has some serious environmental problems, including threats to the survival of native species, which rightly concern most Australians. We have no alternative but to depend upon biological and physical scientists to keep us reliably informed of developments and the growth of knowledge that are relevant to the preservation of a pleasing and fruitful environment.

It is unquestionable that there is much that needs to be done to ensure that continued material improvement in our way of life does not impose unacceptable environmental costs in loss of amenity and extinction of species.

The histories of most societies, including our own, are replete with examples of environmental degradation; from destruction of forests, erosion of top soil, overgrazing, salination, excessive fertilisation, overcropping and extinction of species, to the pollution of rivers, oceans and atmosphere.

What is not sufficiently realised, however, is how frequently the causes of such degradation are to be found in misguided government policies – especially policies which effectively, if not deliberately, undermine or ignore the incentives for good stewardship that firm property rights encourage, or which erect perverse incentives for short term exploitation and damage. Such things, for example, as heavily subsidised irrigation which provides large quantities of water at less than its true cost, thereby encouraging excessive use; subsidisation of dairy production in marginal dairy country; subsidisation of particular crops, leading to monoculture, over-fertilisation, over-production, and neglect of good farming practice; forms of leasehold that create incentives for short-term exploitation rather than long-term husbandry of land and water; and more. Such perverse environmental consequences continue to be encouraged by government policies to this day. Private domestic and industrial action also plays its part, but mostly in the form of the 'externalities' which we have already discussed and whose elimination presents one of the prime challenges for innovative solutions, consistent with the maximum employment of property rights incentives.

In *The Price of Preservation,* Chisholm and Moran (1993:167-180) spell out ten property rights and market-based principles which, if accepted and implemented, would very largely counter and remedy those failures of conservation attributable to government mismanagement and policy shortcomings – especially in not defining and extending property rights and in not providing incentives for private conservation initiatives. What they do in those pages is, in effect, to describe concretely some of the ways in which self-organising processes can be made to work when the law unleashes them, when government limits itself in scope, and when it makes the costs of protection transparent. Their ten principles are (1993:169):

1. Recognise existing property rights and remove barriers to creation of new ones.
2. Unbundle different conservation and resource rights.
3. Allow different conservation and resource rights to be traded.
4. Compensate owners unwilling to sacrifice private goods for conservation.
5. Maximise the ecological value of biodiversity, subject to a ceiling on government spending.
6. Limit the function of government conservation departments to policy co-ordination, and advice on where the allocated funds could be directed.
7. Re-assign public reserve management to private conservation groups, public conservation trusts and local councils under a

variety of licensing arrangements.

8. Encourage conservation on private land through subsidising private conservation initiatives such as voluntary covenants.
9. Establish a conservation budget that reflects community valuations.
10. Make public funding for conservation transparent and subject to scrutiny.

This is an important contribution to the task of devising a framework to guide environmental and biodiversity policies that will preserve liberty and flexibility while routinely dealing with existing environmental problems that entail more than technical or engineering solutions, and those of a kind that are not yet imaginable; but which, in order to be solved, will require the steady acquisition of new knowledge and the continuous use of knowledge widely disseminated throughout the community. These are questions of an order of complexity which cannot be managed from the centre in command-and-control fashion, because no centre could ever effectively collect, let alone digest and use, the information necessary; or encompass the millions of dispersed decisions that might be required.

We are now re-visiting the earlier discussion of the role of property rights, rules of law, prices and markets, in coordinating economic activity, and the Hayekian insight that complex, self-organising systems and discovery processes, such as market orders, may come into existence, paradoxically enough, when the rules governing the order are simple rather than complex. When the actors in such orders respond and coordinate themselves under the guidance of a few simple, easily understood rules, the resulting order may be both extraordinarily complex and smooth-running. Market democracies are one example. The basic rules are that we can act, speak and exchange freely provided we do our neighbours no harm. Democracies, like ecosystems, are complex orders that no individual or group could establish, or maintain, on the basis of specific instructions to each member of the order. Action, in order to achieve the necessary complexity and coordination, has to be decentralised under the control of members responding, in terms of the rules (or natural laws), to their immediate environments, the information available to them, and their personal contingencies. The coherence that emerges nevertheless contains an immense variety of decisions taken by individuals in ignorance of each other.

On the other hand, instrumental organisations – such as government bureaucracies or large firms – are necessarily hierarchical and mechanical. The individual members respond not to a few rules common to all, but to specific rules differentially made to establish roles pieced together to achieve a single end or set of goals. Performance of the roles has to be constantly and hierarchically monitored, and the information necessary to do so must be fed up the hierarchy by operatives at various levels, with many opportunities for information failure or inadequacy. So while instrumental organisations may be a necessary and integral part of the pursuit of limited individual or group objectives within a democratic or market order or ecosystem, the overall order or system could not mimic the instrumental form of organisation and survive.

There is nothing more certain than that the environmental problems we are going to face in the future will increase both in number and complexity. No single, central plan, or an instrumental organisation to implement it, could hope to deal adequately with the thousands or millions of rapidly changing situations requiring all sorts of information and decisions to handle them. What will be required is a set of procedures that will allow the maximum scope for decentralised initiatives following a few simple rules in order to achieve environmental outcomes couched in *generalised* terms; just as the democratic injunction not to harm one's neighbour is a general rule whose actual interpretation in an infinite variety of situations has slowly been established in common law cases.

Our dilemma today is to find the rules for 'sustainable development'. We know what the rules are for development and the production of abundance. We must find rules for sustaining the natural world that are compatible with the rules for creating abundance; bearing also in mind that the rules of human associations are rarely, if ever, absolute. Promises, we say, should always be kept unless they are *coerced* promises. Speech should be free except when it threatens the peace or reputations; and so on. So rules begin as conditional guides to action, subject to modification in the light of experience.

So, where might we begin with the principles of environmental sustainability?

The first imperative is to limit the involvement of politics to the determination of the general laws of environmental protection and then to hand over to the courts the responsibility for the steady evolution of 'environmental common law' that reconciles the principles with the contingencies of experience.

The second imperative is to give long warnings, to avoid sudden and massive disruptions to existing activities, and adequately to compensate those who will incur unavoidable losses. Predictability and certainty are of the essence in encouraging the emergence of a self-ordering system of environmental sustainability.

Against that background, I believe the mainstream environmental ethic can be realised in terms of principles which:

(i) forbid activities which threaten extinction of defined species, except where not doing so would prejudice human lives, health or well-being;

(ii) forbid the use of renewable resources in ways which would threaten their future availability;

(iii) provide incentives, by way of taxation or licences, to avoid polluting activities or the continued use of technologies generating negative externalities;

(iv) to avoid perverse incentives to misuse or overuse renewable resources (e.g. subsidised irrigation leading to salination) or non-renewable resources (e.g. leasehold systems that encourage exploitation of topsoil).

Given these as the fundamental objectives of species protection and environmental conservation law, the next step is to begin operationalising them, and that task, as I indicated above, has been well begun by Chisholm's and Moran's ten principles.

CHAPTER 12

Conclusion: Environmental Ethics & Western Civilisation

In the earlier parts of this study, I identified two main sets of interests and values concerned with environmental protection and species preservation: the 'anthropocentric' or self-interested ethic – taken in the widest sense to include protection and preservation on spiritual or aesthetic grounds – and the 'intrinsic value' and 'sacred earth' ethic. I concluded that the intrinsic value argument is empirically untenable and reduces to the view that some people (probably the majority) value other species and natural processes for their own sakes, simply for the pleasure or satisfaction that observing them or contemplating their existence might give. This, ultimately, is just a version of the mankind-focused or self-interested view. The 'sacred earth' ethic has much in common with the intrinsic value view, but, as a matter of faith, is not open to challenge on empirical grounds.

Essentially, the anthropocentric ethic is utilitarian in arguing that it is useful, or serves various interests of mankind, to preserve species and protect the environment. Mankind must make use of other species and natural objects in all sorts of ways in order to meet a variety of human needs – from the practical matters of food and clothing to the innumerable resources and materials required by a complex and creative culture. The environment and other species are also useful for aesthetic and spiritual purposes – not as things to be transformed, but to be enjoyed simply as they are.

I argued that the 'mainstream' ethic is anthropocentric and utilitarian in adopting the common formula of 'sustainable development' as a shorthand way of integrating the practical and spiritual elements of the ethic. But the thrust of radical or 'deep green' environmentalism goes simultaneously in two directions. The first is to issue a severe warning: If Western societies continue along the path of industrial expansion and more consumption, the effects will be so profound and devastating as to threaten human life and the enjoyment of life. To save ourselves from this fate, we must revolutionise our values and way of life. And to achieve the necessary institutional changes, an ideology of statist intervention, already to hand, must be incorporated. The second direction is to reinforce this basic message with a spiritual adjunct by adopting some features of the intrinsic value and sacred earth views.

But, in the end, the power and persuasiveness of this overall thrust depends crucially upon the scientific truth of the claimed global environmental crisis that is said to be engulfing us and the effectiveness, if it is happening, of the statist, command-and-control remedies that are being offered us.

Our conclusions, on examining the evidence for 'global crisis', and the terms in which it is being presented, were that the claims are misleading and grossly overstated, and that the manner of advancing them revealed a deep-seated mode of propaganda and deception, combined with political manipulation, that was as much ideologically motivated as concerned with environmental protection for its own sake. In addition, the deep green record of predictions and the level of scientific competence in offering them is lamentable.

In 1968, biologist Dr Paul Ehrlich told us: 'We face a very real crisis this instant'. The crisis was the threat of world-wide starvation, and 'famines will be upon us full-scale in 1975'. He told the American people they were a 'cancer on the planet' (Ferguson 1990:29-31). The reality, of course, is that since that prediction was made the world as a whole, a world more populous than ever, has never been so well fed. Where starvation is to be found, the causes do not lie in the incapacity of the earth to feed its people. It was the same Paul Ehrlich who warned in 1980 that the earth was running out of mineral resources. He and a group of colleagues bet economist Julian Simon, in October 1980, that the real (inflation-adjusted) prices of copper, tin, chrome, nickel and tungsten would inevitably rise by 1990; Simon predicted they wouldn't rise in real terms. Simon collected. By 1990, in a world with 800 million more people, the prices of each metal had fallen.

But the propaganda value of apocalyptics and doom-saying, and the media attention it brings, are not the only motives involved. At the root of the ideological energy at work in deep green environmentalism is a strong animus against industrialism, technological innovation, and Western capitalism and its supporting ethos – an ethos drawing upon substantial elements of traditional Christianity, the scientific enlightenment, and traditions oriented towards constantly improving the conditions of human life, under the rule of law.

One of the accusations levelled against a way of life so oriented is that man is alienated from nature and is forced, systematically, to regard nature as a mere commodity, to be used and exploited.

But such a charge dissolves after a moment's reflection. Western achievements depend greatly upon a hard-won, intense and sustained engagement with nature – as a world to be known and understood, as a world to be guarded against in its blacker moods and deceits, and as a world to be cherished and enjoyed. Sheer curiosity, the key motive of science, has been one of the conditions of reaching understandings and forms of cultivation that have contributed immeasurably to the quality and dignity of human life and also to the possibility of learning how to deal with nature in a conserving way. Such achievements are not offences against either human nature or nature herself; they are fulfilments of both. The transformation of its environment, in one way or another, is the law of life for every species, and no less so for mankind. Such transformations, in themselves, imply neither good nor evil. The asteroid which might one day destroy the earth could not be denounced as evil any more than rabbits which destroy pastures. Good and evil apply only to the moral relationships among men, and human transformations of nature can be judged good or bad only in terms of their consequences for *human* lives.

There can be no environmental ethic other than an anthropocentric one because there is only one subject of moral enquiry and discourse and that is the conduct of men and women to each other and the consequences of that conduct for men and women. Environmental values are the subjective sentiments that men and women wish to have taken to account when action with environmental implications is undertaken. How varying values are to be reconciled becomes an issue that only human institutions can resolve, and the latter sections of this study are an attempt to point the way to some long-term solutions, or at least to the principles upon which solutions might be based.

But the animating ethic behind the search for solutions has to be, necessarily, in the opinion of the writer, an enterprising and experimental one based in those traditions of Western civilisation which have so spectacularly improved the conditions of human life and which, when turned to concentrate on conservation of the environment, hold more promise of success than any alternatives of which we have knowledge.

Bibliography & References

Ames, Bruce & Thomas Jukes 1993, Special Books Section, *Reason*, December: 36-7.

Australian Conservation Foundation 1992, *Biodiversity*, Sydney.

Balling, Robert 1993, *Free Perspective* 8(2).

Bandow, Doug 1993, *Ecology as Religion: Faith in Place of Fact*, Competitive Enterprise Institute, Washington D.C.

Bate, Roger & Julian Morris 1994, *Global Warming: Apocalypse or Hot Air?*, The Environment Unit, Institute of Economic Affairs, London.

Beckerman, W. 1990, *Pricing for Pollution*, 2nd Ed., Institute of Economic Affairs, London.

Brandt, R. 1983, The Concept of Moral Right and its Functions', *Journal of Philosophy* 80: 29-45.

Brennan, A. 1984, 'The Moral Standing of Natural Objects', *Environmental Ethics* 6: 35-56.

Brubaker, E. 1975, 'Free Ride, Free Revelation, or Golden Rule', *The Journal of Law and Economics* 18 (1): 147-219.

Brundtland Commission 1990, *Our Common Future*, World Commission on Environment and Development/Oxford University Press, Melbourne.

Buchanan, J. & G. Tullock 1975, 'Polluters' Profits and Political Response: Direct Control Versus Taxes', *The American Economic Review* 65: 139-147.

Callicott, J. Baird 1985, 'Intrinsic Value, Quantum Theory, and Environmental Ethics', *Environmental Ethics* 7(3): 257-275.

Carruthers, Peter 1992, *The Animals Issue: Moral Theory in Practice*, Cambridge University Press, Cambridge.

Carson, Rachel 1963, *Silent Spring*, Hamish Hamilton, London.

Cato Institute 1994, 'Re-cycling is a Waste', *Policy Analysis Report*, Washington, D.C., January 26.

Chisholm, A.G. & A.J. Moran (eds) 1993, *The Price of Preservation*, Tasman Institute, Melbourne.

Clark, S. 1978, 'The Rights of Wild Things', *Inquiry* 22: 171-188.

Coddington, Deborah 1993, 'Little Green Lies', *North and South*, April:85-93.

Cohen, Barry 1992, 'Killers at Large', *The Australian Magazine*, 28-29 March:9-11.

Department of the Arts, Sport, Environment and Tourism 1991, *Biolinks* 1, September.

Diamond, Cora 1978, 'Eating Meat and Eating People', *Philosophy* 53: 465-480.

diZerega, Gus 1992, 'Social Ecology, Deep Ecology, and Liberalism', *Critical Review* 6 (2-3): 305-370.

Ehrenfeld, D. 1988, 'Why Put a Value on Biodiversity', in E. Wilson (ed.) *Biodiversity*, National Academy Press, Washington D.C.

Ehrlich, P.& E Wilson 1991, 'Biodiversity Studies: Science and Policy', *Science* 253: 758-762.

Elder, David 1994, 'God and the Greens', *Environmental Backgrounder* 18, Institute of Public Affairs, 24 January.

Ferguson, Andrew 1990, 'Apocalypse Whenever', *Reason*, April:29-31.

Foundation for Research on Economics and the Environment 1993, *Free Perspective* 3(2), Summer.

Fox, M. 1977-1978, "Animal Liberation": A Critique', *Ethics* 88: 106-118.

Frankel, O.H.& M.E. Soule 1981, *Conservation and Evolution*, Cambridge University Press, Cambridge.

Gill, Peter 1994, 'Greenhouse gas won't go away', *Financial Review*, 28 March.

Global 2000 Report to the President 1980, Vols.1-3, US. Government Printing Office, Washington D.C.

Goklany, Indur M. & Merritt W. Sprague 1992, 'An Alternative Approach to Sustainable Development: Conserving Forests, Habitat and Biological Diversity By Increasing the Efficiency and Productivity of Land Utilization', Office of Program Analysis, US. Department of the Interior, Washington, D.C., July.

Gordon, H. 1954, 'The Economic Theory of a Common- Property Resource: The Fishery', *The Journal of Political Economy* 62: 124-142.

Gore, Al 1992, *Earth in the Balance*, Earth Scan, United Kingdom.

Granberg-Michaelson, Wesley 1993, *Redeeming the Creation - The Rio Earth Summit: Challenges for the Churches*, World Council of Churches, Geneva.

Greenpeace 1992, *Action on Solutions*, Sydney.

Hardin, G. 1968, 'The Tragedy of the Commons', *Science* 162 (December): 1243-1248.

Harlow, E. 1992, 'The Human Face of Nature: Environmental Values and the Limits of Nonanthropocentrism', *Environmental Ethics* 14(l): 27-42.

Hayek, F. A. 1978, 'Competition as a Discovery Procedure', in *New Studies in Philosophy, Politics, Economics and the History of Ideas*, University of Chicago Press, Chicago.

Hooper, Narelle 1994, 'Industry Shapes Up For Fight With Kelly', *Business Review Weekly* 21 February:30-32.

Jastrow, Robert 1992, 'What Happened to the Greenhouse Effect?', *Heritage Lecture* 415, Heritage Foundation, Washington D.C., 31 July.

Knight, F. 1924, 'Some Fallacies in the Interpretation of Social Cost', *The Quarterly Journal of Economics* 38: 582-606

Leone, R. & J. Jackson 1981, 'The Political Economy of Federal Regulatory Activity: The Case of Water Pollution Controls', in G. Fromm (ed.) *Studies in Public Regulation*, MIT Press, Cambridge, Massachusetts.

Lewis, Russell 1992, 'Marketing the Environment', *Economic Affairs*, November: 4-5.

Lieberman, Ben 1994, *Update*, Competitive Enterprise Institute, Washington D.C., January.

Lomasky, L. 1987, *Persons, Rights, and the Moral Community*, Oxford University Press, New York.

Mackie, John 1977, *Ethics: Inventing Right and Wrong*, Penguin Books, Harmondsworth.

Macquarie Dictionary 1981, Macquarie Library Pty. Ltd., Sydney.

Maloney, M. & R McCormick 1982, 'A Positive Theory of Environmental Quality Regulation', *Journal of Law and Economics* 25: 99-123

Mann, Charles C. 1991, 'Extinction: Are Ecologists Crying Wolf?', *Science* 25 (16 August): 736-38.

Meadows, D. et al 1972, *The Limits to Growth: A Report for the Club of Rome's Project on the Predicament of Mankind*, Potomac Associates, New York.

Minogue, Kenneth 1985, *Alien Powers: The Pure Theory of Ideology*, Weidenfeld and Nicolson, London.

Myers, Norman 1979, *The Sinking Ark*, Pergamon, New York.

Nash, R. 1989, *The Rights of Nature: A History of Environmental Ethics*, The University of Wisconsin Press, Wisconsin.

New South Wales Farmers' Association 1993, 'A Rational Approach to Preserving Biodiversity and Endangered Species', *The Primary Report*, 11 October: 1-4.

Norton, B. 1988, 'Commodity, Amenity, and Morality', in E. Wilson (ed.) *Biodiversity*, National Academy Press, Washington D.C.

O'Brien, Brian 1993, *Nationalising the Australian Environment: The Agreements of '92*, Policy Paper No. 23, Institute of Public Affairs, April.

O'Neill, Graeme 1993, 'Going, Going...Found', *Time*, 19 July:48-9.

Pashigian, B. 1984, 'The Effect of Environmental Regulation on Optimal Plant Size and Factor Shares', *Journal of Law and Economics* 27: 1-28.

——— 1985, 'Environmental Regulation: Whose Self-Interests are Being Protected', *Economic Inquiry* 13: 551-584.

Passmore, John 1974, *Man's Responsibility for Nature*, Charles Scribner's Sons, New York.

Postrel, Virginia 1990, 'The Green Road to Serfdom', *Reason*, April: 22-28.

Rachels, J. 1990, *Created From Animals: The Moral Implications of Darwinism*, Oxford University Press, Oxford.

Regan, Tom 1982, *All That Dwell Therein: Animal Rights and Environmental Ethics*, University of California Press, Berkeley and Los Angeles.

——— 1984, *The Case for Animal Rights*, Routledge, London.

Ridley, M. 1988, 'Privatising America's West', *The Economist* 309, 22 October: 21-24.

Rodd, R. 1990, *Biology, Ethics, and Animals*, Oxford University Press, Oxford.

Rolston, Holmes 1988, *Environmental Ethics*, Temple University Press, Philadelphia.

Salmon, Jeffrey 1992, 'PC Science', *Commentary*, October: 54-6.

Shaw, J. 1988, 'The Vanishing Jungle: Ecologists make friends with the economists', *The Economist* 309, 15 October: 25-28.

Simon, Julian & Aaron Wildavsky 1993, *Assessing the Empirical Basis of the 'Biodiversity Crisis'*, Competitive Enterprise Institute, Washington, D.C., May.

Smart, J. & B. Williams 1973, *Utilitarianism: For and Against*, Cambridge University Press, Cambridge.

Smith, Adam 1982 [1790], *The Theory of Moral Sentiments*, Liberty Classics, Indianapolis IN.

Smith, R. 1988, 'Private Solutions to Conservation Problems', in T. Cowen (ed) *The Theory of Market Failure: A Critical Examination*, The Cato Institute, George Mason University Press, VA.

Singer, Peter 1975, *Animal Liberation*, Jonathan Cape, London.

——— 1979, *Practical Ethics*, Routledge, London.

——— 1993, *How Are We To Live: Ethics in an Age of Self-Interest*, Text Publishing, Melbourne.

Stavins, Robert 1993, 'Comments on 'Lethal Model 2: The Limits to Growth Revisited', by William Nordhaus', in *Brookings Papers on Economic Activity*.

Steffen, L. 1992, 'In Defense of Dominion', *Environmental Ethics* 14, Spring: 63-80.

Stone, C. 1972, *Should Trees have Standing? Towards Legal Rights for Natural Objects*, Avon Books, New York.

Sydney Morning Herald 1994, 'Researchers question value of lead removal strategies', 21 February.

Taylor, P. 1986, *Respect for Nature: A Theory of Environmental Ethics*, Princeton University Press, New Jersey.

Wamsley, John 1993, *The Sanctuary Movement in Australia*, Stirling, South Australia.

Westra, L. 1989, 'Ecology and Animals: Is There a Joint Ethic of Respect?', *Environmental Ethics*, 11 (Fall):215- 230.

Whitmore, T. C. & J. A. Sayer (eds.) 1992, *Tropical Deforestation and Species Extinction*, Chapman and Hall, London.

Wilson, E. (ed.) 1988, *Biodiversity*, National Academy Press, Washington D.C.

——— 1989, 'Threats to Biodiversity', *Scientific American*, September: 60-66.

——— 1994, 'Death Wish: Guess Which Species is Heading for Extinction…Ours!', *Good Weekend Magazine, Sydney Morning Herald*, 12 February.

Wilson, James Q. 1993, *The Moral Sense*, The Free Press, New York.

Index